Texts and Monographs in Computer Science

Suad Alagić and Michael A. Arbib
The Design of Well-Structured and Correct Programs
1978. x, 292pp. 14 illus. cloth

Michael A. Arbib, A. J. Kfoury, and Robert N. Moll
A Basis for Theoretical Computer Science
1981. vii, 220 pp. 49 illus. cloth

F. L. Bauer and H. Wössner
Algorithmic Language and Program Development
1982. xvi, 497pp. 109 illus. cloth

Kaare Christian
The Guide to Modula-2
1985. xx, 437pp. 46 illus. cloth

Edsger W. Dijkstra
Selected Writings on Computing: A Personal Perspective
1982. xvii, 362pp. 1 illus. cloth

Nissim Francez
Fairness
1986. x, 176pp. 147 illus. cloth

Peter W. Frey, Ed.
Chess Skill in Man and Machine, 2nd Edition
1983. xiv, 329 pp. 104 illus. cloth

R. T. Gregory and E. V. Krishnamurthy
Methods and Applications of Error-Free Computation
1984. xii, 189pp. 1 illus. cloth

David Gries, Ed.
Programming Methodology: A Collection of Articles by Members of IFIP WG2.3
1978. xiv, 437pp. 68 illus. cloth

David Gries
The Science of Programming
1981. xvi, 366pp. cloth

A. J. Kfoury, Robert N. Moll, and Michael A. Arbib
A Programming Approach to Computability
1982. viii, 251pp. 36 illus. cloth

E. V. Krishnamurthy
Error-Free Polynomial Matrix Computations
1985. xii, 144pp. cloth

Franco P. Preparata and Michael Ian Shamos
Computational Geometry: An Introduction
1985. xii, 376pp. 231 illus. cloth

Brian Randell, Ed.
The Origins of Digital Computers: Selected Papers
3rd Edition. 1982. xvi, 580 pp. 126 illus. cloth

Error-Free Polynomial Matrix Computations

E. V. Krishnamurthy

Springer-Verlag
New York Berlin Heidelberg Tokyo

E. V. Krishnamurthy
Department of Computer Science
University of Waikato
Hamilton
New Zealand

Series Editor

David Gries
Department of Computer Science
Cornell University
Upson Hall
Ithaca, NY 14853
U.S.A.

Library of Congress Cataloging in Publication Data
Krishnamurthy, E. V.
  Error-free polynomial matrix computations.
  (Texts and monographs in computer science)
  Bibliography: p.
  Includes index.
  1. Interpolation—Data processing. 2. Matrices—
Data processing. 3. Polynomials—Data processing.
I. Title. II. Series.
QA281.K75   1985   511'.42   85-2845

ISBN-13: 978-1-4612-9572-3   e-ISBN-13: 978-1-4612-5118-7
DOI: 10.1007/978-1-4612-5118-7

Typeset by The Universities Press (Belfast) Ltd., Northern Ireland.

9 8 7 6 5 4 3 2 1

*In memory of my Parents—
who taught me about icons,
signs, symbols, and
their interpretation.*

Preface

This book is written as an introduction to polynomial matrix computations. It is a companion volume to an earlier book on *Methods and Applications of Error-Free Computation* by R. T. Gregory and myself, published by Springer-Verlag, New York, 1984. This book is intended for seniors and graduate students in computer and system sciences, and mathematics, and for researchers in the fields of computer science, numerical analysis, systems theory, and computer algebra.

Chapter I introduces the basic concepts of abstract algebra, including power series and polynomials. This chapter is essentially meant for bridging the gap between the abstract algebra and polynomial matrix computations.

Chapter II is concerned with the evaluation and interpolation of polynomials. The use of these techniques for exact inversion of polynomial matrices is explained in the light of currently available error-free computation methods.

In Chapter III, the principles and practice of Fourier evaluation and interpolation are described. In particular, the application of error-free discrete Fourier transforms for polynomial matrix computations is considered.

Chapter IV deals with the theory and construction of Hensel Codes of single variable rational polynomials, based on formal power series over the finite fields (Hensel fields). Each rational polynomial is assigned a unique Hensel Code and the arithmetic operations with Hensel Codes as operands are mathematically equivalent (isomorphic) to those same arithmetic operations with the corresponding rational polynomials as operands. Also the interconversions of rational polynomials and their

Hensel Codes are described; in this context, the relationship of Hensel Code to Padé approximants is indicated. The application of Hensel Codes to polynomial matrix inversion is explained; in addition, the Hensel's methods for iterative inversion and generalized inversion of matrices, whose elements are Hensel Codes, are described.

In Chapter V, we discuss briefly the matrix computations over the Euclidean and non-Euclidean domains. The concept of Hensel Code is extended to a multivariable rational polynomial. The applications of Hensel Code for direct and iterative inversion and generalized inversion of multivariable rational polynomial matrices are considered.

Explicit prerequisites for reading this book are abstract and linear algebra, number theory, error-free computation methods based on modular arithmetic, and an exposure to data structures and algorithms related to symbolic and algebraic computation.

I wish to acknowledge with appreciation the contributions of those who in various ways influenced this book, in particular Robert T. Gregory, University of Tennessee, Knoxville, Tennessee, U.S.A.; Azriel Rosenfeld, University of Maryland, College Park, Maryland, U.S.A.; Bruno Buchberger, Johannes Kepler Universität, Linz, Austria; and Reinhardt Klette, Friedrich-Schiller Universität, Jena, DDR.

I am indebted to the Curriculum Development Cell established at the Indian Institute of Science, Bangalore, by the Ministry of Education and Culture, India, for financial assistance toward the preparation of the earlier versions of this manuscript and to the University of Waikato, Hamilton, New Zealand for the facilities provided during the final stages of the manuscript. Also, my heartfelt thanks go to Carol Stanford, who cheerfully and skillfully typed the final manuscript. It is also my pleasant duty to thank the editorial staff of Springer-Verlag for their cooperation and efforts.

Hamilton, New Zealand   E. V. KRISHNAMURTHY
September 1985

Contents

Glossary

A^-	a g-inverse of a matrix A
A^{-1}	the inverse of a matrix A
A_L^-	a least-square g-inverse of a matrix A
A_M^-	a minimum norm g-inverse of a matrix A
A_R^-	a reflexive g-inverse of a matrix A
A^T	transpose of a matrix A
A^+	the Moore–Penrose inverse of A
$A(x)$	a single variable polynomial matrix A
$A(x_1, x_2, \ldots, x_n)$	a multivariable polynomial matrix A
$A \cong B$	A is isomorphic to B or A is equivalent to B
(A, P)	Algebraic system with elements A and operations P
$a \mapsto b$	assignment of a to b
$a \mid b$	a divides b
$[a/b]$	the integer part of the quotient a/b
$\lvert a \rvert_m$	the nonnegative remainder of a when divided by m
$a \equiv_m b$ or $a \equiv \lvert b \rvert_m$ or $a = b \bmod m$	a is congruent to b mod m
$a(x) \bmod m(x)$	remainder when $a(x)$ is divided by $m(x)$
$a(x)^{-1}$	inverse transform of $a(x)$ over a field F
adj A	adjoint of a matrix A
D	integral domain
deg $a(x)$	degree of $a(x)$
det A	determinant of a matrix A
DFT	discrete Fourier transform
F	field

FFT	fast Fourier transform
$\mathbb{F}_N$	the set of order-N Farey fractions
$F[x]$	polynomials over a field F
$F(x)$	rational functions over a field F
$F[[x]]$	Formal Power Series over a field F
$F((x))$	Power series rational functions over the field F
$F\langle x \rangle$	extended formal power series over a field F
$F_p(x)$	set of rational polynomials over a finite field $(\mathbb{I}_p, +, \cdot)$
$F_p((x))$	power series rational function over $(\mathbb{I}_p, +, \cdot)$
$F[x_1, x_2, \ldots, x_n]$	multivariable polynomials over a field F
$F(x_1, x_2, \ldots, x_n)$	multivariable rational polynomials over a field F
$F[[x_1, x_2, \ldots, x_n]]$	multivariable formal power series over a field F
$F((x_1, \ldots, x_n))$	multivariable power series rational functions over a field F
$F\langle x_1, \ldots, x_n \rangle$	multivariable extended power series over a field F
$F_p(x_1, \ldots, x_n)$	multivariable rational polynomials over a finite field $(\mathbb{I}_p, +, \cdot)$
$F_p((x_1, \ldots, x_n))$	Power series rational functions over a finite field $(\mathbb{I}_p, +, \cdot)$
G	group
$\gcd(a, b)$	greatest common divisor of a and b
$H(p, r, \alpha)$	the Hensel Code of a rational number α
$H(p, r, \alpha(x))$	the Hensel Code of a rational polynomial $\alpha(x)$ over a finite field $(\mathbb{I}_p, +, \cdot)$
$H(p, r, \alpha(x_1, \ldots, x_n))$	the Hensel Code of a multivariable rational polynomial $\alpha(x_1, \ldots, x_n)$ over a finite field $(\mathbb{I}_p, +, \cdot)$
$\mathbb{I}$	the set of integers
$\mathbb{I}[x]$	polynomials over $\mathbb{I}$
$\mathbb{I}[x_1, \ldots, x_n]$	multivariable polynomials over $\mathbb{I}$
$\mathbb{I}/(m)$	quotient ring of integers modulo m
$\mathbb{I}_p$	residue class of integers modulo p $\{0, 1, 2, \ldots, p-1\}$
$\mathbb{I}_p[x]$	polynomials over a finite field $(\mathbb{I}_p, +, \cdot)$
$\mathbb{I}_p[[x]]$	power series over a finite field $(\mathbb{I}_p, +, \cdot)$
$\mathbb{I}_p[x_1, \ldots, x_n]$	multivariable polynomials over a finite field $(\mathbb{I}_p, +, \cdot)$
$\mathbb{I}_p[[x_1, \ldots, x_n]]$	multivariable power series over a finite field $(\mathbb{I}_p, +, \cdot)$
$\text{lcm}(a, b)$	least common multiple of a and b.
$\mathbb{N}$	the set of nonnegative integers
NTT	number-theoretic transform
$O(r)$	truncation of terms having degree $\geq r$
$\text{ord } a(x)$	order of $a(x)$
$P(L/M, F_p(x))$	Padé rational polynomials in x over a finite field $(\mathbb{I}_p, +, \cdot)$ with numerator degree $\leq L$ and denominator degree $\leq M$
$P(L/M, N, x)$	Padé rational polynomials in x over integers with coefficient size $\leq N$ and numerator degree $\leq L$, and denominator degree $\leq M$
$P(L_1, L_2, \ldots, L_n/$ $M_1, M_2, \ldots, M_n,$ $F_p(x_1, \ldots, x_n))$	multivariable Padé rational polynomials over a finite field $(\mathbb{I}_p, +, \cdot)$ where degree of $x_i \leq L_i$ in numerator and degree of $x_i \leq M_i$ in denominator
$P(L_1, L_2, \ldots, L_n/$ $M_1, M_2, \ldots, M_n,$ $N, x_1, \ldots, x_n)$	multivariable Padé rational polynomials over integers with coefficient size $\leq N$ and degree of $x_i \leq L_i$ in numerator and degree of $x_i \leq M_i$ in denominator

$\mathbb{Q}$	the set of rational numbers
$Q(D)$	field of quotients of D
$\mathbb{Q}_p$	the set of p-adic numbers
$\mathbb{Q}(x)$	rational polynomials over the field of rationals $\mathbb{Q}$
$\mathbb{Q}(x_1, x_2, \ldots, x_n)$	multivariable rational polynomials over the field of rationals $\mathbb{Q}$
R	ring
$R[x]$	polynomials over a ring R
$R[x]_{m(x)}$	ring of polynomials modulo $m(x)$
$R[x]/m(x)$	quotient ring of polynomials modulo $m(x)$
$R[[x]]$	power series over a ring R
$R[[x]]_m$	ring of truncated power series order m
$R[[x]]/(x^m)$	quotient ring of power series modulo x^m
$R[x_1, x_2, \ldots, x_n]$	ring of polynomials in n variables $x_1, \ldots, x_n$
$\mathbb{S}_m$	set of symmetric residues, $\{-(m-1)/2, \ldots, -1, 0, 1, \ldots, (m-1)/2\}$
$v(A)$	column vectorization of an $(m \times n)$ matrix A by concatenating all its m rows end to end with the first row on the left and the last row on the right and transposing it to a column vector $(mn \times 1)$
$\in$	belongs to
$\notin$	does not belong to
$\phi(\cdot)$	norm
$\phi(m)$	Euler's totient function
$\displaystyle\prod_{i=1}^{n}$	product symbol
$\displaystyle\sum_{i=1}^{n}$	summation symbol
ω	primitive root
$\otimes$	Tensor or Kronecker product of matrices
$*$	convolution

CHAPTER I
Algebraic Concepts

1 Introduction

This introductory chapter on algebraic concepts is meant for bridging the gap between the abstract algebra and the computational polynomial matrix algebra. Much of this background material which follows is abstract algebra and we shall assume that the reader is familiar with these basic ideas. Most of these results are available in standard text books on algebra and related subjects: Albert [1956], Berlekamp [1968], Childs [1979], Jacobson [1976], Lipson [1981], Zariski and Samuel [1975]. Here we shall quickly review these results and in doing so, we shall state many theorems without proof. The reader is directed to the above texts for detailed proofs.

2 Groups, Rings, Integral Domains, and Fields

2.1 Groups

A set G of elements $a, b, c, \ldots$ is said to form a group $(G, \cdot)$ with respect to a binary operation, say multiplication (denoted by $\cdot$), if for every $a, b, c \in G$

(i)   the product $a \cdot b$ is in G (closure);

(ii) associative law $a \cdot (b \cdot c) = (a \cdot b) \cdot c$ holds;

(iii) there exist solutions x and y in G of the equations $a \cdot x = b$ and $y \cdot a = b$;

(iv) if the product $a \cdot b = b \cdot a$, we call G commutative or abelian;

(v) every group G contains a unique element e called *identity* such that for $a \in G$, $a \cdot e = e \cdot a = a$. Moreover, every $a \in G$ has an inverse element b such that $a \cdot b = e = b \cdot a$. We denote $b = a^{-1}$.

Then the solution of equations in (iii) can be written as $x = a^{-1} \cdot b$, $y = b \cdot a^{-1}$.

2.1.1 EXAMPLE. Consider the set $\mathbb{I}_p$ of integers modulo p, where p is a prime integer. In $\mathbb{I}_p$, addition (+) is performed as in $\mathbb{I}$, the set of integers, except that all the numbers are replaced by their remainders after division by p (denoted by modulo p or mod p or $\|_p$). Note that $(\mathbb{I}_p, +)$ is an abelian group with 0 as identity. Also if a is any element in the set $\{0, 1, \ldots, p-1\}$ then $(p-a)$ is its additive inverse.

2.2 Rings

A ring $(R, +, \times)$ is a set closed under two binary operations $+, \times$ such that $(R, +)$ is an abelian group, $\times$ is associative and has an identity (denoted by 1) with respect to $\times$ and satisfies for every $a, b, c \in R$

$$a \times (b + c) = (a \times b) + (a \times c)$$
$$(a + b) \times c = (a \times c) + (b \times c)$$

called the distributive law. We denote the identity element with respect to $+$ by 0 and the inverse of a with respect to $+$ by $-a$.

2.2.1 *Commutative.* A ring is called commutative if the product $a \times b = b \times a$ for all $a, b \in R$.

2.2.2 *Zero divisor.* An element $a \neq 0$ in R is said to be a zero divisor, if $a \times b = 0$ for some $b \neq 0$ in R (symmetrically b is also a zero divisor).

2.2.3 *Unit.* An element $u \neq 0$ in R is said to be a unit, if it is invertible: $u \times v = 1$ for some $v \in R$ and we denote this by $v = u^{-1}$.

2.2.4 *Division ring.* A ring R is called a division ring R^*, if $R^* = (R - \{0\})$ is a group with respect to $\times$.

2.2.5 EXAMPLE. Consider $\mathbb{I}_m$ the set of integers modulo m, where m is a positive integer. In $\mathbb{I}_m$, addition (+) and multiplication ($\times$) are performed as in $\mathbb{I}$, the set of integers, but all the numbers y are replaced by their remainders after division by m (y mod m). This is a commutative ring with the additive identity 0 and multiplicative identity 1. In particular, consider $\mathbb{I}_6$. We have $2 \times 3 = 0$ and so $\mathbb{I}_6$ is a ring with zero divisors. Also 1 and 5 are units, since these are invertible: $1^{-1} = 1$, $5^{-1} = 5$.

2.3 Integral domain

An integral domain D is a commutative ring with no zero divisors: $a \times b = 0$ implies $a = 0$ or $b = 0$. This also implies that for all a, b, c in D, $a \times b = a \times c$ and $a \neq 0$ implies $b = c$ or the cancellation law holds.

2.3.1 *Divisibility*. In D we say a is divisible by b (or b divides a denoted by $b \mid a$) if there exists an element c in D such that $a = b \times c$. We also say a is a multiple of b and b is a divisor of a.

2.3.2 *Unit*. If u in D divides its unity element 1, then u is called a unit of D. u is a unit of D, if it has an inverse in D. The inverse of a unit is also a unit.

2.3.3 *Associates*. Two elements $a, b \in D^*$, where $D^* = (D - \{0\})$ are called associates, if $a = u \times b$, where u is a unit in D. This property of association is an equivalence relation on D^*.

2.3.4 *Divisors*. A divisor a of b is called a proper divisor of b, if a is neither a unit nor an associate of b.

2.3.5 *Prime*. A quantity p of an integral domain is called a prime or irreducible quantity of D if $p \neq 0$ is not a unit of D and the only divisors of p in D are units and associates of p. Every associate of a prime is a prime.

2.3.6 *Composite*. A composite quantity of D is an $a \neq 0$ which is neither a prime nor a unit of D.

2.3.7 gcd. For $a, b \in D$, an element $c \in D$ is called a greatest common divisor (gcd) of a and b, if $c \mid a$ and $c \mid b$ and c is a multiple of every other element which divides both a and b.

2.3.8 lcm. For $a, b \in D$, an element $c \in D$ is called a least common multiple (lcm) of a and b if $a \mid c$ and $b \mid c$ and c is a divisor of every other element which is a multiple of both a and b.

2.3.9 *Relative prime*. We call two elements $a, b \in D$ relatively prime if $\gcd(a, b) = 1$.

2.3.10 EXAMPLE. The set of integers $\mathbb{I}$ under addition (+) and multiplication ($\times$) is an integral domain.

(i)   The units are 1 and -1;
(ii)   3 is a gcd of 9 and 12;
(iii, -3 also a gcd of 9 and 12;
(iv)   3 and -3 are associates;
(v)   3 and 14 are relatively prime;
(vi)   19 and -19 are primes;
(vii) 18 is a composite.

It is now natural to ask whether or not every composite a of D may be written as $a = p_1 \times p_2 \times \cdots \times p_k$ for a finite number of primes k of D.

2.4 Unique factorization domain

An integral domain D is called a unique factorization domain, if for all $a \in D^*$, either a is a unit or else a can be expressed as a finite product of primes (viz. $a = p_1 \times p_2 \times \cdots \times p_k$) such that this factorization into primes is unique up to associates and reordering. (i.e., if $a = p_1 \times p_2 \times \cdots \times p_k$ and $a = q_1 \times q_2 \times \cdots \times q_m$, where p_i $(1 \leq i \leq k)$ and q_j $(1 \leq j \leq m)$ are primes, then $k = m$ and there exists a reordering of the q_j's such that p_i is an associate of q_i $(1 \leq i \leq k)$.

2.4.1 Theorem. *If D is a unique factorization domain and if $a, b \in D$ are not both zero, then $\gcd(a,b)$ exists and is unique.*

The above theorem guarantees the existence and uniqueness of gcd.

2.4.2 *Choice of unique* gcd *in D.* In D, two elements c and d are associates if and only if the product $c \times u$ (denoted hereafter by $cu) = d$ for some unit u. If c is a gcd of a and b then so is any associate $d = cu$ and conversely if c and d are gcd's of a and b then c must be an associate of d.

In order to make gcd unique, we need to impose an additional condition. Since the property of being an associate (or association relationship) is an equivalence relation, it partitions D^* into equivalence classes of associated elements for any element $a \in D^*$. If we denote this class by $[a] = \{ua : u \in U(D)\}$, where $U(D) =$ group of units of D, we can pick up a distinguished representative in this class by using a convention. Let us denote any element

$$a = \Delta(a)u(a),$$

where $\Delta(a)$ denotes the distinguished representative of the associate class containing a and $u(a)$ the corresponding unit. Then we can choose $\Delta(a)$ as the unique distinguished representative.

2.4.3 EXAMPLES.

(i) In $\mathbb{I}$ we have

$$\Delta(a) = |a|, \, u(a) = \mathrm{sgn}(a),$$

where

$$\mathrm{sgn}(a) = \begin{cases} -1, & a < 0 \\ +1, & a \geq 0. \end{cases}$$

(ii) In any general integral domain D, we choose unity as the distinguished representative of $[u]$, where u is a unit.

2.4.4 *Uniqueness of* lcm. The lcm of two elements $a, b \in D$, when it exists, can be made by choosing a unique gcd since

$$\text{lcm}(a,b) = \frac{ab}{\gcd(a,b)}.$$

2.5 Euclidean domains

A Euclidean domain is an integral domain D together with a valuation (also called Euclidean degree function) $v : D - \{0\} \rightarrow \mathbb{N}$ (the set of non-negative integers) having the following properties:

(i)   For all $a, b \in D - \{0\}$, $v(ab) \geq v(a)$.
(ii) For all $a, b \in D$ with $b \neq 0$, there exist elements $q, r \in D$ such that $a = bq + r$, where either $r = 0$ or $v(r) < v(b)$.
(iii) Any Euclidean domain is an integral domain in which every pair of elements (a,b), not both zero, has a gcd g which is expressible as

$$g = sa + tb \, (s, t \in D).$$

2.5.1 EXAMPLE. In $\mathbb{I}$, $\gcd(8,20) = 4$; we have $4 = s(8) + t(20)$ with $s = -2$, and $t = 1$.

2.5.2 *Extended Euclidean Algorithm for* gcd (EEA). We now recall the Extended Euclidean Algorithm (Section 6 of Chapter 1, Gregory and Krishnamurthy [1984]) for finding $g = \gcd(a,b)$ and compute $s, t \in D$ such that $g = sa + tb$. (For a detailed discussion see Knuth [1981], Lipson [1981].)

We will first consider the case where $D = \mathbb{I}$, the set of integers. To find a pair of integers s and t which satisfy

$$sa + tb = g$$

we use the Euclid's division algorithm successively, as follows:

$$\begin{aligned}
\text{Let} \quad r_1 &= a + b(-q_1) \\
r_2 &= b + r_1(-q_2) \\
&= a(-q_2) + b(1 + q_1 q_2) \\
r_3 &= r_1 + r_2(-q_3) \\
&= a(1 + q_2 q_3) + b(-q_1 - q_3 - q_1 q_2 q_3) \\
&\;\;\vdots \\
r_n &= r_{n-2} + r_{n-1}(-q_n) \\
&= sa + tb \\
0 &= r_{n-1} + r_n(-q_{n+1}) \\
&= xa + yb.
\end{aligned}$$

The above computation can be expressed in the form of a table:

		a	1	0
		b	0	1
q_1	r_1	1		$-q_1$
q_2	r_2	$-q_2$		$1+q_1q_2$
q_3	r_3	$1+q_2q_3$		$-q_1-q_3-q_1q_2q_3$
$\vdots$	$\vdots$	$\vdots$		$\vdots$
q_n	r_n	s		t
q_{n+1}	0	x		y

The above table can be thought of as a sequence of row vectors generated by a matrix–vector product. Starting with

$$\begin{bmatrix} a & 1 & 0 \\ b & 0 & 1 \end{bmatrix}$$

and $[1, -q_1]$ we obtain

$$[r_1, 1, -q_1] = [1, -q_1]\begin{bmatrix} a & 1 & 0 \\ b & 0 & 1 \end{bmatrix}$$

$$[r_2, -q_2, 1+q_1q_2] = [1, -q_2]\begin{bmatrix} b & 0 & 1 \\ r_1 & 1 & -q_1 \end{bmatrix}$$

and so on. In these equations

$$q_1 = \left[\frac{a}{b}\right]$$

$$q_2 = \left[\frac{b}{r_1}\right]$$

$$\vdots$$

$$q_{n+1} = \left[\frac{r_{n-1}}{r_n}\right].$$

where [ ] denotes the integer part of the quotient.

2.5.3 **Remark.** In order to obtain the unique gcd, over a general integral domain D, however, we need to replace a and b by their distinguished representative of the associate class. Accordingly s and t are to be suitably modified finally, as in the following algorithm.

2.5.4 *Algorithm* EEA *over a general D.*

    begin

        comment Given a and b in a Euclidean domain D,
        compute $g = \gcd(a,b)$ and find $s, t \in D$ such that $sa + tb = g$;

$$A \leftarrow \Delta(a); \qquad A0 \leftarrow \Delta(b);$$
$$B \leftarrow 1; \qquad B0 \leftarrow 0;$$
$$C \leftarrow 0; \qquad C0 \leftarrow 1;$$
while $A0 \neq 0$ **do**
  begin
    $Q \leftarrow$ quotient $(A, A0);$
    $RA \leftarrow A - Q \times A0;$
    $RB \leftarrow B - Q \times B0;$
    $RC \leftarrow C - Q \times C0;$
    $A \leftarrow A0; \qquad A0 \leftarrow RA;$
    $B \leftarrow B0; \qquad B0 \leftarrow RB;$
    $C \leftarrow C0; \qquad C0 \leftarrow RC;$
  end;
  $g \leftarrow \Delta(A); s \leftarrow u^{-1}(a) \times u^{-1}(A) \times B;$
  $t \leftarrow u^{-1}(b) \times u^{-1}(A) \times C$

end.

2.5.5 EXAMPLE. We will illustrate this algorithm by finding the gcd(a,b) where $D = \mathbb{I}$ with $a = 25$, $b = 15$

Q	25(A) 15($A0$)	1(B) 0($B0$)	0(C) 1($C0$)
1	10(RA)	1(RB)	-1(RC)
1	5	-1	2
2	0	3	-5

Thus gcd$(25,15) = 5$

$$u^{-1}(a) = 1$$
$$u^{-1}(b) = 1$$
$$u^{-1}(A) = 1$$
$$\Delta(A) = 5$$
$$s = -1$$
$$t = 2.$$

Also

$$0 = 3 \times 25 - 5 \times 15$$
$$5 = -1 \times 25 + 2 \times 15.$$

2.6 Field F

A field F is a nontrivial commutative ring in which every nonzero element is a unit. Also F is a Euclidean domain where the remainder on division is always zero. F is also a unique factorization domain, but a nonzero element has no prime factorization as there are no primes in F.

2.6.1 EXAMPLES.

(i) The finite commutative ring $(\mathbb{I}_p, +, \times)$ (where p is a prime having a finite number of elements $\{0, 1, \ldots, (p-1)\}$, $+$ is addition modulo p, and $\times$ is multiplication modulo p) is a finite field.

(ii) The set of rational numbers $\mathbb{Q}$ form a field.

2.7 Use of EEA to compute multiplicative inverse

The following theorem enables us to compute $b^{-1}(m)$, the multiplicative inverse of b modulo m, in a finite commutative ring or in a finite field using the Euclidean algorithm 2.5.4.

2.7.1 **Theorem.** If $\gcd(m,b) = 1$ and if $1 = sm + tb$ then $b^{-1}(m) = |t|_m$.

PROOF. Since $1 = sm + tb$ we can write

$$1 = |sm + tb|_m = |tb|_m$$
$$= |t|b|_m|_m$$

and so $b^{-1}(m) = |t|_m$. □

2.7.2 EXAMPLE. Find the multiplicative inverse of 12 mod 19. Since $\gcd(12,19) = 1$ we know that this inverse exists.

We obtain the following table:

		19	1	0
		12	0	1
1	7	1	-1
1	5	-1	2
1	2	2	-3
2	1	-5	8
2	0	12	-19

Thus $|12^{-1}|_{19} = 8$ or $|12 \cdot 8|_{19} = 1$.

2.7.3 **Remark.** Note that the computation of the second column (enclosed in dashed lines) in the above table is redundant, as we need only t.

This means that it is sufficient to start with the seed matrix $\begin{bmatrix} m & 0 \\ b & 1 \end{bmatrix}$ to obtain $b^{-1} \bmod m$.

2.8 Additional properties of EEA

To describe some additional properties of EEA, we reformulate the steps using the following notation.

Step 1. Choose the seed matrix:

$$\begin{bmatrix} A & B \\ A_0 & B_0 \end{bmatrix} = \begin{bmatrix} a & b \\ c & d \end{bmatrix},$$

where a and c are given elements and b and d are arbitrary elements over D.

Step 2. For $i = 1, 2, \ldots, n+1$, while $A_{i-1} \neq 0$ determine the quotient q_i and the remainder A_i in the division of A_{i-2} by A_{i-1}. Then $A_i = A_{i-2} - q_i A_{i-1}$.

Step 3. Likewise, define

$$B_i = B_{i-2} - q_i B_{i-1}.$$

Step 4. Terminate when $A_{n+1} = 0$. At this point $A_n = \gcd(a,c)$.

2.8.1 **Theorem.** *In the above formulation of the EEA if we chose $a = m$, $b = 0$, and $0 < c < m$ such that $\gcd(c,m) = 1$, then for $i = 1, 2, \ldots, n$*

$$\left| \frac{B_i}{A_i} \right|_m = \left| \frac{d}{c} \right|_m.$$

PROOF. Since $a = m$ and $b = 0$, we have $A_i d \equiv B_i c \pmod{m}$. Consequently, since

$$\left| \frac{u}{v} \right|_m \equiv \left| \frac{r}{s} \right|_m$$

if and only if

$$us \equiv vr \pmod{m}$$

the conclusion follows.   □

2.8.2 **Theorem.** *Given any two elements r and s in the finite ring $(\mathbb{I}_m, +, \times)$, where $\gcd(s,m) = 1$, the EEA seeded with the matrix*

$$\begin{bmatrix} m & 0 \\ s & r \end{bmatrix}$$

will terminate (for some $n+1$ such that $A_{n+1}=0$). At this point

$$\left|\frac{r}{s}\right|_m = |B_n|_m.$$

PROOF. Since $A_n = \gcd(a,c) = \gcd(s,m) = 1$ it follows from Theorem 2.8.1 that

$$\left|\frac{B_n}{A_n}\right|_m = |B_n|_m = \left|\frac{r}{s}\right|_m. \qquad \square$$

2.8.3 EXAMPLE. If we wish to compute $|\frac{10}{13}|_{125}$ we can use the seed matrix

$$\begin{bmatrix} 125 & 0 \\ 13 & 10 \end{bmatrix}$$

and obtain the table

	125	0
	13	10
9	8	−90
1	5	100
1	3	−190
1	2	290
1	1	−480
2	0	1250

Thus $|\frac{10}{13}|_{125} = |-480|_{125} = 20$.

2.9 Ideals, principal ideals, quotient rings

An ideal L in a ring R is a nonempty subset of R such that

(i) $a, b \in L$ implies $a + (-b) \in L$ ($-b$ denotes the inverse of b with respect to $+$);

(ii) $a \in L, r \in R$ implies $ar, ra \in L$.

Then L consists of zero alone and will be called the zero ideal of R, or L may be seen to be a subring of R with the property that the products ar, ra of any element $a \in L$ and any element $r \in R$ are in L.

If R is a commutative ring and $m \in R$, then the set of all ring multiples of m, $(m) = \{km : k \in R\}$ is (i) an ideal of R and (ii) the smallest ideal of R containing m. This ideal (m) is called the principal ideal because it is generated by a single element m.

2.9.1 *Quotient rings.* Let L be an ideal in a commutative ring R. Then the collection $R/L = \{a + L : a \in R\}$ of additive cosets of L in R is a quotient ring under the operations:

$$(a+L)+(b+L)=(a+b)+L; \quad -(a+L)=-a+L;$$
$$0_{R/L}=0+L=L; \quad (a+L)(b+L)=ab+L;$$
$$1_{R/L}=1+L.$$

Also R/L is a homomorphic image of R under the natural map $M:a \mapsto a+L$ and R/L is a commutative ring.

2.9.2 EXAMPLE. Since (m) is an ideal in the set of integers $\mathbb{I}$, $\mathbb{I}/(m)=\{a+(m):a \in \mathbb{I}\}$ under the coset operation defined in 2.9.1 is a ring (quotient ring of integers mod m); also it is a homomorphic image of $\mathbb{I}$ under the map

$$M:a \mapsto a+(m).$$

Thus $a+(m)=b+(m)$ if and only if $a \equiv_m b$; where $\equiv_m$ denotes mod m.

3 Power Series and Polynomials

Let R be a commutative ring. We now define a formal power series over R.

3.1 Formal power series $R[[x]]$

We define $R[[x]]$ as the set of all expressions with infinite number of terms of the form

(3.1) $$a(x)=a_0+a_1x+a_2x^2+\cdots,$$

where the coefficients a_i ($i \in \mathbb{N}$, the set of nonnegative integers) of $a(x)$ lie in R. We also write

(3.2) $$a(x)=\sum_{i=0}^{\infty} a_i x^i.$$

We regard $a(x)$ as a formal expression in the sense $\sum$ in (3.2) does not represent a sum of values in R; hence convergence is not implied. Each such $a(x)$ is called a formal power series over R in an indeterminate x.

Equality of formal power series is defined as

$$a(x)=b(x) \text{ implies } a_i=b_i, \quad \text{for all } i \in \mathbb{N}.$$

Each $a_i x^i$ is called a term (of degree i) of $a(x)$. If $a_i \neq 0$, we say $a_i x^i$ occurs in $a(x)$. We also denote $(-a_i)x^i$ by $-a_i x^i$, $1 \cdot x^i$ by x^i, ax^1 by ax, and ax^0 by a. We further eliminate terms with zero coefficients. Thus an element $a \in R$ is represented by

$$a = a + 0 \cdot x + 0 \cdot x^2 + \cdots$$

and hence $a \in R[[x]]$.

Also similarly

$$x = 0 \cdot + 1 \cdot x + 0 \cdot x^2 + \cdots \in R[[x]].$$

It is easily seen that $R[[x]]$ is a ring under the operation $+$, $-$, $\cdot$ (with $0 = 0(x)$ and $1 = 1(x)$) to be defined below:

$$c(x) = a(x) + b(x) \quad \text{if and only if } c_i = a_i + b_i \quad (i = 0, 1, 2, \ldots)$$

$$= a(x) \cdot b(x) \quad \text{if and only if } c_i = \sum_{j=0}^{i} a_j b_{i-j}$$

$$= -a(x) \quad \text{if and only if } c_i = -a_i$$

$$= 0 \quad \text{if and only if } c_i = 0$$

$$= 1 \quad \text{if and only if} \begin{cases} c_i = 1, & \text{for } i = 0 \\ c_i = 0, & \text{otherwise,} \end{cases}$$

where the right hand side binary operations on the coefficients are those of R.

We define the degree of $a(x)$ in (3.2) [denoted by $\deg a(x)$] thus

$$\deg a(x) = -\infty \quad \text{if } a(x) = 0$$

$$= n \quad \text{if for some } n \in \mathbb{N}, \, a_n \neq 0 \quad \text{and } a_i = 0, \quad \text{for } i > n,$$

$$= +\infty \quad \text{otherwise.}$$

3.2 Polynomial rings

We now define $R[x]$, as the set of all polynomials over R in an indeterminate x by

$$R[x] = \{a(x) \in R[[x]] : \deg a(x) < \infty\}.$$

If $a(x) \in R[x]$ has $\deg a(x) = n$ then we write

(3.3)
$$a(x) = \sum_{i=0}^{n} a_i x^i.$$

We observe, $R[x]$ under the formal power series operations $+$, $-$, $\cdot$, 0, 1 is a subring of $R[[x]]$. We call $R[x]$ the polynomial ring over R in an indeterminate x.

In (3.3) if $a_n \neq 0$, we can write

$$a(x) = a_n x^n + \text{lower degree terms.}$$

a_0 is called the constant term and a_n is called the leading degree coefficient. We also say

$$\deg 0 = -\infty.$$

Also we call $a(x)$ monic, if $a_n = 1$.

For power series (3.2) we introduce the notion of order (ord $a(x)$)

$$\text{ord } a(x) = n \quad \text{if } a(x) = a_n x^n + \text{higher degree terms}$$

$$\text{ord } 0 = +\infty.$$

Here a_n is called the low order coefficient and a_0 is called the constant term.

From the definitions of $\deg a(x)$ and ord $a(x)$ we have:

(i)   In $R[x]$:

$$\deg(a(x) \cdot b(x)) \leqslant \deg a(x) + \deg b(x)$$

$$\deg(a(x) + b(x)) \leqslant \max\{\deg a(x), \deg b(x)\};$$

(ii) In $R[[x]]$:

$$\text{ord}(a(x) \cdot b(x)) \geqslant \text{ord } a(x) + \text{ord } b(x)$$

$$\text{ord}(a(x) + b(x)) \geqslant \min[\text{ord } a(x), \text{ord } b(x)].$$

Based on the above concepts we can analogously define the formal power series and polynomials over D and F. These results are given in the following theorems.

3.2.1 **Theorem.** *If D is an integral domain, then so are the set of polynomials $D[x]$ and the power series $D[[x]]$ over D.*

3.2.2 **Theorem.** *$a(x)$ is a unit of $D[x]$ if and only if $a(x)$ is a unit of D.*

3.2.3 **Theorem.** *The degree function in $D[x]$ and order function in $D[[x]]$ obey*

$$\deg(a(x) \cdot b(x)) = \deg a(x) + \deg b(x)$$

$$\text{ord}(a(x) \cdot b(x)) = \text{ord } a(x) + \text{ord } b(x).$$

3.2.4 **Theorem.** *The units of $F[x]$ are the nonzero constant polynomials of F.*

3.2.5 **Theorem.** *$F[x]$ is an integral domain.*

3.2.6 **Theorem.** *$a(x) = a_0 + a_1 x + \cdots$ is a unit in $D[[x]]$ if and only if a_0 is a unit in D.*

PROOF (*If part*). If $a(x) \cdot b(x) = 1$ for $b(x) = b_0 + b_1 x + \cdots \in D[[x]]$ then equating the coefficients gives $a_0 b_0 = 1$ or a_0 is a unit in D.

(*Only if part*). Since $b(x) = b_0 + b_1 x + \cdots \in D[[x]]$ satisfying $a(x) \cdot b(x) = 1$, by equating the coefficients we obtain

$$a_0 b_0 = 1$$
$$a_0 b_1 + a_1 b_0 = 0$$
$$a_0 b_n + a_1 b_{n-1} + \cdots + a_n b_0 = 0.$$

Since

$$b_0 = a_0^{-1}$$
$$b_1 = -a_0^{-1}(a_1 b_0)$$
$$b_n = -a_0^{-1}(a_1 b_{n-1} + \cdots + a_n b_0),$$

$b(x)$ exists and satisfies

$$a(x)b(x) = 1 \quad \text{or} \quad b(x) = a(x)^{-1}. \qquad \square$$

We call $b(x)$ the inverse transform of $a(x)$ over $D[[x]]$.

3.2.7 Corollary. $a(x) = a_0 + a_1 x + \cdots$ *is a unit in* $F[[x]]$, *where* F *is a field, if and only if* $a_0 \neq 0$.

3.2.8 Theorem. $F[[x]]$ *is still not a field and is only an integral domain. In particular,* $x, x^2, \ldots, x^i$ *are not units in* $F[[x]]$. *To facilitate this extension we define the extended formal power series.*

3.2.9 *Extended power series* $F\langle x \rangle$. Let

$$(3.4) \qquad\qquad a(x) = \sum_{i=-\infty}^{\infty} a_i x^i,$$

where $a_i \neq 0$ for only finitely many $i < 0$. That is to say that for every $a(x) \in F\langle x \rangle$ there is an integer n (which depends on $a(x)$) such that $a_i = 0$, for all $i < n$.

Thus we may rewrite (3.4) as

$$(3.5) \qquad\qquad a(x) = \sum_{i=n}^{\infty} a_i x^i \qquad (n \in \mathbb{I}).$$

3.2.10 Theorem. *Let* $a_i \in F$; *then the set* $F\langle x \rangle$ *of all extended formal power series is a field.*

PROOF. Let $a(x) \in F\langle x \rangle$, $a(x) \neq 0$. We first prove $a(x)$ is a unit in $F\langle x \rangle$. Then there exists an $n \in \mathbb{I}$ such that $a_i = 0$ for $i < n$. Let n be the greatest such integer, i.e., $a_i = 0$ for $i < n$ and $a_n \neq 0$ (since $a(x) \neq 0$, such an integer exists).

We then have

$$a(x) = a_n x^n + a_{n+1} x^{n+1} + \cdots$$
$$= x^n (a_n + a_{n+1} x + \cdots)$$
$$= x^n b(x),$$

where $b(x) = b_0 + b_1 x + b_2 x^2 + \cdots$ is an element of $F[[x]]$ and has $b_0 = a_n \neq 0$. Note that $b(x)$ is a unit in $F[[x]]$ and hence a unit in $F\langle x \rangle$. Thus $a(x)^{-1} = x^{-n} \cdot b(x)^{-1}$. $\square$

3.2.11 *The quotient ring* $R[x]/(m(x))$. Let $m(x)$ be a monic polynomial in $R[x]$, where R is commutative and degree $m(x) > 0$. Then $m(x)$ is an ideal in $R[x]$, and $R[x]/m(x)$ is a ring known as the quotient ring.

3.2.12 *The quotient ring* $R[[x]]/(x^m)$. Let $R[[x]]$ be a power series ring and m a fixed positive integer. We then take the ideal (x^m) and form the quotient ring $R[[x]]/(x^m)$ (the quotient ring of power series mod x^m). The congruence relation $\equiv_{x^m}$ induced by (x^m) satisfies

$$a(x) \equiv_{x^m} b(x) \qquad \text{if and only if } a(x) - b(x) \in (x^m),$$
$$\text{if and only if } a_i = b_i \ (i = 0, 1, \ldots, m-1).$$

Hence $a(x), b(x) \in R[[x]]$ are congruent mod x^m whenever they agree in their first m terms.

3.3 Field of quotients

Given an integral domain D we can construct a field F that includes D. Thus in F, every nonzero element in D will be invertible. The following theorem tells us how to construct the field F from the integral domain D.

3.3.1 Theorem. *Let D be an integral domain. Define a relation $\simeq$ on $D \times D^* (D^* = D - \{0\})$ according to*

(3.6) $(a/b) \simeq (c/d)$ *if and only if $ad = bc$.*

Then (3.6) is an equivalence relation on $D \times D^*$.
   Let $Q(D) = D \times D^*/\simeq$ and denote the equivalence class $[(a,b)]$ by the fraction a/b. Thus according to (3.6) we have

$$a/b = c/d \Leftrightarrow ad = bc.$$

Define $+$ and $\cdot$ in $Q(D)$ by

$$a/b + c/d = (ad + bc)/bd$$
$$a/b \cdot c/d = (a \cdot c)/(b \cdot d).$$

Then $Q(D)$ is a field in which

$$0 = 0/1, \qquad 1 = 1/1;$$
$$-a/b = (-a)/b; \quad and$$
$$(a/b)^{-1} = b/a \ (a \neq 0).$$

3.3.2 EXAMPLES.
(i)   If $D = \mathbb{I}$, then the quotient field of the set of integers $\mathbb{I}$, is the field of rational numbers, denoted by $\mathbb{Q}$.
(ii) Since $F[x]$ is an integral domain, it has a field of quotients which we denote by $F(x)$:

$$F(x) = \{a(x)/b(x) : a(x), b(x) \in F[x], b(x) \neq 0\};$$

$F(x)$ is called the field of rational functions over F.
(iii) $F[[x]]$ is also an integral domain and so we can construct a field of quotients, denoted by $F((x))$:

$$F((x)) = \{a(x)/b(x) : a(x), b(x) \in F[[x]], b(x) \neq 0\}.$$

$F((x))$ is called a power series rational function.

3.4 Examples

3.4.1 *Inversion.* Let

$$a(x) = a_0 + a_1 x + a_2 x^2 + a_3 x^3 \in F[x],$$

where

$$F = (\mathbb{I}_p, +, \cdot).$$

Let $a(x)^{-1} = b(x) = b_0 + b_1 x + b_2 x^2 + \cdots + b_n x^n + \cdots$. Then we have

$$a(x) \cdot a(x)^{-1} = 1 + 0 \cdot x + 0 \cdot x^2 + \cdots 0 \cdot x^n + \cdots.$$

Thus (using Theorem 3.2.6)

$$a_0 \cdot b_0 = 1$$
$$a_0 \cdot b_1 + a_1 b_0 = 0$$
$$\vdots$$
$$a_0 b_n + a_1 b_{n-1} + \cdots + a_n b_0 = 0,$$

and so

$$b_0 = a_0^{-1} \bmod p$$
$$b_1 = -a_0^{-1}(a_1 b_0)$$
$$\vdots$$
$$b_n = -a_0^{-1}(a_1 b_{n-1} + a_2 b_{n-2} + \cdots + a_n b_0)$$
$$= -a_0^{-1}\left(\sum_{i=1}^{n} a_i b_{n-i}\right)$$

3.4.2 *Inversion.* Let

$$a(x) = 4 + 2x + 3x^2 \in \mathbb{I}_7[x]$$
$$a(x)^{-1} = b_0 + b_1 x + b_2 x^2 + \cdots.$$

Using the formula in Example 3.4.1 above we get

$$b_0 = 4^{-1} = 2$$
$$b_1 = -2(2.2) \bmod 7 = 6$$
$$b_2 = -2[2.6 + 3.2] = 6,$$

or $a(x)^{-1} = 2 + 6x + 6x^2 + \cdots$.

3.4.3 **Remark.** The determination of $a(x)^{-1}$ can also be carried out by solving the linear system

$$\begin{bmatrix} a_0 & 0 & 0 & \cdots & 0 \\ a_1 & a_0 & 0 & \cdots & 0 \\ a_2 & a_1 & a_0 & \cdots & 0 \\ \vdots & \vdots & \vdots & & \vdots \\ a_n & a_{n-1} & a_{n-2} & \cdots & a_0 \end{bmatrix} \begin{bmatrix} b_0 \\ b_1 \\ b_2 \\ \vdots \\ b_n \end{bmatrix} = \begin{bmatrix} 1 \\ 0 \\ 0 \\ \vdots \\ 0 \end{bmatrix}$$

or

$$Ab = e_1.$$

where the right side vector is the unit vector $e_1 = (1, 0, \ldots, 0)^T$ and the unknown vector

$$b^T = (b_0, b_1, \ldots, b_n)^T.$$

The matrix A is lower triangular and since a_0^{-1} exists, A is nonsingular and hence b^T is unique. Note also that given b^T, the coefficients a_i can be got by successive back substitution.

3.4.4 *Inversion.* Let

$$a(x) = 6 + 2x + 3x^2 \in \mathbb{I}_{11}[x]$$
$$a(x)^{-1} = 2 + 3x + 9x^2 + \cdots.$$

3.4.5 *Truncation of a formal power series.* Let $a(x) = 1 + 2x \in F[x]$, where $F = (\mathbb{I}_7, +, \cdot)$. Then

$$a(x)^{-1} = 1 + 5x + 4x^2 + 6x^3 + 2x^4 + 3x^5$$
$$+ x^6 + 5x^7 + 4x^8 + 6x^9 + 2x^{10} + 3x^{11} + \cdots.$$

Then

$$a(x)^{-1} \bmod x^5 = 1 + 5x + 4x^2 + 6x^3 + 2x^4.$$

3.4.6 **Remark.** As already stated in Theorems 3.2.1 and 3.2.5, both $D[x]$ and $F[x]$ are integral domains. In particular, $F[x]$ is an Euclidean domain (Section 2.5) as will be seen below.

3.5 Division property of $F[x]$

Let

$$a(x) = a_m x^m + \cdots + a_0 \quad \text{and} \quad b(x) = b_n x^n + \cdots + b_0$$

be polynomials in $F[x]$ and let $b_n \neq 0$. Then there exists a unique quotient $q(x) \in F[x]$ and remainder $r(x) \in F[x]$ that satisfy the Euclidean division property (Section 2.5):

$$a(x) = b(x)q(x) + r(x)$$

with

$$v(r(x)) = \deg r(x) < v(b(x)) = \deg b(x).$$

3.5.1 **Remainder Theorem.** *If we use the division property* 3.5 *to write*

$$a(x) = (x - c)q(x) + r(x),$$

where $b(x) = x - c$, $(c \in F)$, *is of degree one and* $r(x) = r = constant \in F$, *then*

$$r = a(c).$$

Also, if $a(x)$, $b(x)$, $q(x)$, *and* $r(x) \in F[x]$ *are polynomials satisfying the division property, then for* $c \in F$,

$$a(c) = b(c)q(c) + r(c).$$

3.5.2 **Factor Theorem.** *The division algorithm and remainder theorem imply the factor theorem:* α *is a root of* $f(x)$ *if and only if* $(x - \alpha)$ *divides* $f(x)$ *(denoted by* $(x - \alpha) \mid f(x)$*).*

3.5.3 *Associates.* In Section 2.3.3 we defined associates in an integral domain. We call two nonzero polynomials $a(x)$ and $b(x)$ associated if $a(x) \mid b(x)$ and $b(x) \mid a(x)$. Then

$$a(x) = q(x)b(x), \ b(x) = p(x)a(x);$$

thus, $a(x) = q(x)p(x)a(x)$ or $q(x)p(x) = 1$, where $q(x)$ and $p(x)$ are nonzero constants. Thus $a(x)$ and $b(x)$ are associated if and only if each is a nonzero multiple of each other. Also this implies that if $a(x)$ and $b(x)$ are associated and monic (coefficient of the highest degree term is unity) they must be equal.

3.5.4 gcd. We define the gcd of polynomials $f_1(x), \ldots, f_t(x)$, not all zero, to be any monic polynomial $g(x)$ which divides all the $f_i(x)$, $(i = 1, 2, \ldots, t)$, and is such that if $d(x)$ divides every $f_i(x)$ then $d(x)$ divides $g(x)$.

If $h(x)$ is a second such polynomial, then $g(x)$ and $h(x)$ divide each other; or $g(x)$ and $h(x)$ are associated monic polynomials and so they are equal. Thus the choice of gcd is unique.

3.5.5 Theorem. *Let $a(x)$ and $b(x)$ be polynomials not both zero. Then there exist polynomials $s(x)$ and $t(x)$ such that*

$$g(x) = a(x)s(x) + b(x)t(x)$$

is a monic common divisor of $a(x)$ and $b(x)$; moreover, $g(x)$ is then the unique gcd of $a(x)$ and $b(x)$; (this property of Euclidean domain is already stated in Section 2.5).

3.5.6 *Relative primeness.* If $a(x)$ and $b(x)$ are relatively prime then (see Section 2.3.9)

$$a(x)s(x) + b(x)t(x) = 1.$$

Note: $s(x)$ and $t(x)$ need not be unique.

3.5.7 Theorem. *Let $a(x) \neq 0$ and $b(x) \neq 0$ have respective degrees m and n. Then polynomials $s(x) \neq 0$, of degree at most $(n-1)$ and $t(x) \neq 0$ of degree at most $(m-1)$ such that $a(x)s(x) + b(x)t(x) = 0$, exist if and only if $a(x)$ and $b(x)$ are not relatively prime.*

3.5.8 Remark. Since both integers $\mathbb{I}$ and polynomials $F[x]$ belong to Euclidean domain, the extended Euclidean algorithm (EEA) 2.5.4 can be directly used for finding the gcd of two polynomials belonging to $F[x]$, by defining the Euclidean valuation function suitably, and using polynomial arithmetic in $F[x]$. In this case, we make the choice of gcd unique by choosing the monic polynomial as the distinguished representative.

3.5.9 Remarks. Since the ring $\mathbb{I}_m$ of integers modulo m, $R[x]_{m(x)}$ of polynomials mod $m(x)$ and $R[[x]]_m$ of truncated power series mod x^m are all isomorphic to quotient rings $-\mathbb{I}/(m)$, $R[x]/(m(x))$ and $R[[x]]/(x^m)$ respectively, we can use the algorithm described in Section 2.7 to compute the multiplicative inverse $b(x)$ of a polynomial $a(x)$ such that $a(x) \cdot b(x) \bmod x^r = 1$.

Similarly, we can directly apply the Theorem 2.8.2 to obtain the truncated power series $a(x) \cdot b^{-1}(x) \bmod x^r$.

3.6 Examples

3.6.1 *Division of polynomials.* Find the quotient $q(x)$ and remainder $r(x)$ in the division of

$$a(x) = x^4 \in F[x], \qquad F = (\mathbb{I}_7, +, \cdot)$$
$$b(x) = (1 + x) \in F[x], \qquad F = (\mathbb{I}_7, +, \cdot).$$

We use the long division scheme (by representing only the coefficients corresponding to decreasing powers) as below:

		1,	6,	1,	6
1,	1	1,	0,	0,	0,	0
		1,	1,	0,	0,	0
			6,	0,	0,	0
			6,	6,	0,	0
				1,	0,	0
				1,	1,	0
					6,	0
					6,	6
						1

Thus, $q(x) = x^3 + 6x^2 + x + 6$; $r(x) = 1$.

3.6.2 *Greatest common divisor.* We now use Algorithm 2.5.4 to find the gcd of two polynomials over $F = (\mathbb{I}_7, +, \cdot)$. Let

$$a(x) = x^2 + 5x + 6;$$
$$b(x) = x^2 + 3x + 2.$$

We use the seed matrix

$$\begin{bmatrix} x^2 + 5x + 6 & 0 \\ x^2 + 3x + 2 & 1 \end{bmatrix}.$$

The computation is displayed in the following table, where each $q_i(x)$ is the quotient

$q_i(x)$	$x^2 + 5x + 6$	0
	$x^2 + 3x + 2$	1
1	$2x + 4$	-1
$(4x + 4)$	0	$(4x + 5)$

Thus a greatest common divisor is

$$\gcd(x^2 + 5x + 6, \, x^2 + 3x + 2) = 2x + 4.$$

3.6.3 **Remark.** Note that any nonzero element in a field F is a unit of the domain of polynomials over that field; so any nonzero multiple of a greatest common divisor is also a greatest common divisor (over F). It is therefore conventional (as mentioned earlier) to divide a gcd by its leading coefficient so as to make it monic and unique. Using the convention, we obtain by dividing by 2

$$\gcd(x^2+5x+6, x^2+3x+2) = x+2.$$

3.6.4 *Inversion of a polynomial using gcd algorithm.* We can use the analogue of Theorem 2.8.2 to obtain the truncated formal power series (the truncation being modulo x^r) of the inverse of a polynomial $a(x)$ which is relatively prime to x^r. For this purpose we use the seed matrix

$$\begin{bmatrix} x^r & 0 \\ a(x) & 1 \end{bmatrix}.$$

Let $x^r = x^4$ and $a(x) = 1+2x$ over $(\mathbb{I}_7, +, \cdot)$.

The computation is displayed in the following table.

	x^4	0
	$1+2x$	1
$4x^3+5x^2+x+3$	4	$3x^3+2x^2+6x+4$
$4x+2$	0	$2x^4$

As mentioned in Remark 3.6.3, since any nonzero element in F is a unit of the domain of polynomials over F, the remainder obtained at the last but one step in column 1 need not necessarily be the identity element 1. It could be a unit c in F. Accordingly the result obtained $b(x)$ (say) would be such that $b(x) \cdot a(x) = c \in F$. This has to be multiplied by the multiplicative inverse c^{-1} of the unit c in the field F to obtain the unique inverse $a(x)^{-1} \bmod x^4$. In the above example, we thus have

$$a(x)^{-1} \bmod x^4 = 4^{-1} \cdot (3x^3+2x^2+6x+4)$$
$$= 1+5x+4x^2+6x^3$$

(compare this result with that obtained in Example 3.4.5 above).

3.6.5 *gcd using algorithm* EEA. We will now illustrate Algorithm 2.5.4 for finding $\gcd(2x^2+3x+5, 3x^2+2x+6)$ over $(\mathbb{I}_7, +, \cdot)$; this is similar to Example 2.5.5. Here

$$a(x) = 2x^2+3x+5 = (x^2+5x+6) \cdot 2$$
$$b(x) = 3x^2+2x+6 = (x^2+3x+2) \cdot 3.$$

Thus

$$\Delta a = x^2+5x+6, \qquad u(a) = 2$$
$$\Delta b = x^2+3x+2, \qquad u(b) = 3.$$

We obtain the following table:

	$x^2+5x+6(A)$ $x^2+3x+2(A0)$	$1(B)$ $0(B0)$	$0(C)$ $1(C0)$
1	$2x+4(RA)$	$1(RB)$	$-1(RC)$
$4x+4$	0	$3x+3$	$4x+5$

Thus

$$\Delta(A) = x+2, \qquad u(A) = 2$$
$$s = 1 \cdot u^{-1}(a) \cdot u^{-1}(A) = 2$$
$$t = -1 \cdot u^{-1}(b) \cdot u^{-1}(A) = 1.$$

Also

$$x+2 = (2x^2+3x+5) \cdot 2 + (3x^2+2x+6) \cdot 1$$
$$0 = (3x+3)(x^2+5x+6) + (4x+5)(x^2+3x+2).$$

3.7 Congruence properties

The striking similarity between the set of integers $\mathbb{I}$ and $F[x]$, as already stated, classify them under the structure known as Euclidean domains with a suitably defined valuation. This classification also helps us to introduce the congruence properties for polynomials over $F[x]$ in a manner similar to that for $\mathbb{I}$, the set of integers (Gregory and Krishnamurthy [1984]). Let $a(x)$, $A(x)$, $M(x) \in F[x]$. If $M(x)$ divides $A(x) - a(x)$ or in other words, there exists a polynomial $q(x)$ such that

$$A(x) = q(x)M(x) + a(x),$$

then we say $A(x)$ is congruent to $a(x)$ modulo $M(x)$ or

$$A(x) \equiv a(x) \bmod M(x).$$

3.7.1 **Theorem.** *If* $B(x) \equiv b(x) \bmod M(x)$ *and* $A(x) \equiv a(x) \bmod M(x)$, *where* $A(x), B(x) \in F[x]$; *then*

$$A(x) \pm B(x) \equiv (a(x) \pm b(x)) \bmod M(x)$$
$$A(x) \cdot B(x) \equiv (a(x) \cdot b(x)) \bmod M(x)$$

where $\pm, \cdot$ *are operations in* F.

3.7.2 **Theorem.** *If* $a(x)$ *and* $A(x)$ *have both degrees less than* $M(x)$ *and if* $a(x) \equiv A(x) \bmod M(x)$ *then* $A(x) = a(x)$.

PROOF. $a(x) - A(x) = q(x) M(x)$. But no nonzero multiples of $M(x)$ can have degree less than the degree of $M(x)$. Hence any polynomial of degree less than that of $M(x)$ lies in a distinct residue class mod $M(x)$.

□

3.7.3 *Principal ideal domain.* A principal ideal domain is an integral domain in which every ideal (Section 2.9) is principal.

3.7.4 **Theorem.** *A Euclidean domain is a principal ideal domain.*

3.7.5 EXAMPLES.

(i)   The set of integers I is a principal ideal domain in which every ideal has the form (m), $m \geq 0$;
(ii) $F[x]$ is a principal ideal domain in which every ideal $\neq (0)$ has the form $(m(x))$, where $m(x)$ is monic;
(iii) $F[[x]]$ is a principal ideal domain in which every ideal $\neq (0)$ is of the form (x^m) for $m \geq 0$.

3.7.6 *Truncated power series operations.* Let $F[[x]]$ be a power series ring and m is a fixed integer > 0. In the set

$$F[[x]]_m = \{a(x) \in F[[x]] : \deg a(x) < m\}$$

of all truncated power series having degree $< m$, we can define the arithmetic operations $+, -, 0, 1$ as on $R[[x]]$ and the truncation mod x^m operator. By Theorems 3.7.1 and 3.7.2 it is clear that

$$(a(x) \oplus b(x)) \bmod x^m = a(x) \bmod x^m + b(x) \bmod x^m$$

$$(a(x) \odot b(x)) \bmod x^m = a(x) \bmod x^m \cdot b(x) \bmod x^m,$$

where $a(x), b(x) \in F[[x]]$ and the left hand side operations $\oplus$ and $\odot$ are power series operations and the right hand side operations are polynomial operations over F. Also

$$(a(x))^{-1} \bmod x^m = (a(x) \bmod x^m)^{-1} \bmod x^m,$$

where $(a(x))^{-1} \cdot a(x) \equiv 1 \bmod x^m$; in other words inversion and truncation results in the same quantity as truncation and inversion mod x^m.

3.7.7 **Remarks.** The truncated power series operations enable us to carry out the isomorphic image scheme for performing arithmetic operations with rational polynomials, in much the same way as we did for performing rational number arithmetic using the finite segment p-adic computations (see Chapter II, Gregory and Krishnamurthy [1984]). The structural simplicity of I_p compared with the set of rationals Q arises due to the fact that finite field arithmetic is simpler than the rational arithmetic. In using the finite field arithmetic all the values are bounded by p and the time-consuming intermediate gcd calculations to reduce the fractions to lowest terms are not required. A similar situation arises when rational polynomial arithmetic is replaced by finite segment power-series arithmetic over a finite field. This will be described in detail in Chapter IV.

4 Chinese Remainder Theorem and Interpolation

One of the oldest known algorithms for computing solutions to a simultaneous system of linear congruences over the integers makes use of a classical theorem from the theory of numbers, called the Chinese remainder theorem (see Gregory and Krishnamurthy [1984], Lipson [1981], MacDuffee [1948]).

4.1 Theorem (Chinese Remainder Theorem). *Let* $p_0, p_1, \ldots, p_n$ *be* $(n+1)$ *relatively prime integers* > 1 *over* I. *Then for any* $r_0, r_1, \ldots, r_n$ $(r_i < p_i)$ *there is a unique integer* r *satisfying*

$$r < \prod_{i=0}^{n} p_i = M$$

and $r_i \equiv r \bmod p_i$; *the integer* r *can be computed using*

$$r = \sum_{i=0}^{n} \left(\frac{M}{p_i}\right) r_i T_i \bmod M,$$

where T_i *is the solution of*

$$\left(\frac{M}{p_i}\right) T_i \equiv 1 \bmod p_i.$$

4.1.1 Remark. Note that (M/p_i) and p_i are relatively prime and so T_i can be uniquely determined as $(M/p_i)^{-1} \bmod p_i$ using the EEA (Section 2.7).

4.1.2 EXAMPLE. Let $r_1 = 2$, $r_2 = 1$, $r_3 = 3$ with $p_1 = 3$, $p_2 = 4$, $p_3 = 5$. Then

$$M = 60, \qquad \frac{M}{p_1} = 20, \qquad \frac{M}{p_2} = 15, \qquad \frac{M}{p_3} = 12.$$

Thus

$$20T_1 \equiv 1 \bmod 3, \qquad T_1 = 2$$
$$15T_2 \equiv 1 \bmod 4, \qquad T_2 = 3$$
$$12T_3 \equiv 1 \bmod 5, \qquad T_3 = 3$$

and $r = |(20 \cdot 2 \cdot 2 + 15 \cdot 1 \cdot 3 + 12 \cdot 3 \cdot 3)|_{60} = 53$. We are now interested in the generalization of the Chinese remainder Theorem 4.1 to polynomials in $F[x]$ (as well to general Euclidean domains) as stated in the following theorem.

4.2 **Theorem.** *Given the residues* $(r_0, r_1, \ldots, r_n)$ *corresponding to moduli* $(p_0, p_1, \ldots, p_n)$ *which are pairwise relatively prime, there exists an element* r *of the same domain such that* $r \equiv r_i \bmod p_i$ *for* $i = 0, 1, \ldots, n$ *and* r *is uniquely determined, if size* $r < \text{size} \prod_{i=0}^n p_i = M$ *(where size is defined under a suitable valuation; see Sections 2.5 and 3.5). Also*

$$r = \sum_{i=0}^n \left(\frac{M}{p_i}\right) r_i T_i \bmod M,$$

where T_i *is the solution of* $(M/p_i)T_i \equiv 1 \bmod p_i$.

4.2.1 **Remark.** To prove Theorem 4.2 for polynomials in $F[x]$, we go through two stages. In the first stage, we establish the equivalence of the statement of Theorem 4.2 with the generalized Lagrangian interpolation formula used in numerical analysis (Blum [1972]); in the second stage we prove the validity of the latter.

4.3 Equivalence to Lagrange interpolation

Let $r(x)$ be any nth degree polynomial in $F[x]$. Let the values of $r(x)$ at the $(n+1)$ points $x_0, x_1, \ldots, x_n \in F$ respectively be $r_0 = r(x_0)$, $r_1 = r(x_1), \ldots, r_n = r(x_n)$. The Lagrange interpolation (passage from the functional values to the coefficients of $r(x)$) consists in recovering $r(x)$ from the evaluated r_i $(i = 0, 1, \ldots, n)$ using

$$L_n(x) = \sum_{i=0}^n r_i T_i \prod_{\substack{i \neq k \\ 0 \leqslant k \leqslant n}} (x - x_k),$$

where

$$T_i = \frac{1}{\prod_{\substack{i \neq k \\ 0 \leqslant k \leqslant n}} (x_i - x_k)} = \left[\frac{M}{p_i(x)}\right]^{-1} \bmod(x - x_i),$$

where

$$M = \prod_{i=0}^n (x - x_i) = \prod_{i=0}^n p_i(x) \quad \text{and} \quad p_i(x) = (x - x_i).$$

Since we restrict ourselves to linear polynomials $(x - x_i)[i = 0, 1, \ldots, n]$, for a given polynomial $r(x)$ the residue or remainder computation is equivalent to $r(x) \bmod(x - x_i) = r_i$ (by the remainder Theorem 3.5.1); this evaluates $r(x)$ at a point $x_i \in F$. Since the x_i's are distinct, $(x - x_i)$'s are pairwise relatively prime and the problem of recovering $r(x)$ is identical to the use of generalized Chinese remainder Theorem 4.2 in $F[x]$. We now have the following theorem.

4.3.1 **Theorem.** *Given a set of* $(n+1)$ *distinct points (elements of* $F)$

$x_0, x_1, \ldots, x_n$ *there exists a unique polynomial* $L_n(x) \in F[x]$ *having degree* $\leq n$ *such that* $L_n(x_i) = r(x_i)$.

PROOF. The polynomial $L_n(x)$ is such that $L_n(x_i) = r(x_i), 0 \leq i \leq n$. If $s(x)$ is another polynomial of degree $\leq n$, with the property that $L_n(x_i) = s(x_i)$, then $s(x_i) - L_n(x_i) = 0$. Since $s(x) - L_n(x)$ is a polynomial of degree $\leq n$, which vanishes at $(n+1)$ distinct points, it must be identically zero; i.e., $s(x) = L_n(x)$.

Another method of proving this theorem is to set

$$L_n(x) = a_0 + a_1 x + \cdots + a_n x^n$$

and simply write the necessary and sufficient conditions on the coefficients, namely,

$$L_n(x_i) = a_n x_i^n + \cdots + a_0 \qquad (0 \leq i \leq n).$$

This is a system of $(n+1)$ linear equations in the $(n+1)$ unknowns $a_0, \ldots, a_n$ with the coefficient matrix

$$X = \begin{bmatrix} 1 & x_0 & x_0^2 & \cdots & x_0^n \\ 1 & x_1 & x_1^2 & \cdots & x_1^n \\ \cdot & \cdot & \cdot & \cdots & \cdot \\ \cdot & \cdot & \cdot & \cdots & \cdot \\ 1 & x_n & x_n^2 & \cdots & x_n^n \end{bmatrix}.$$

The determinant of X is the Vandermonde determinant, D given by

$$D = \prod_{0 \leq j < i \leq n} (x_i - x_j).$$

Since all the x_i's are distinct over F, $D \neq 0$ and so L_n exists and is unique.

$\square$

4.3.2 EXAMPLE. Consider $F = (\mathbb{I}_7, +, \cdot)$;

$$x_0 = 1, \qquad x_1 = 2, \qquad x_2 = 3 \quad \text{and}$$
$$r(x_0) = 3, \qquad r(x_1) = 2, \qquad r(x_2) = 1.$$

Then

$$r(x) = \frac{3(x-2)(x-3)}{(1-2)(1-3)} + \frac{2(x-1)(x-3)}{(2-1)(2-3)} + \frac{1(x-1)(x-2)}{(3-1)(3-2)}$$

$$= \frac{3}{2}[x^2 - 5x + 6] + \frac{2}{-1}[x^2 - 4x + 3] + \frac{1}{2}[x^2 - 3x + 2]$$

$$= 5(x^2 - 5x + 6) + 5(x^2 - 4x + 3) + 4(x^2 - 3x + 2)$$

$$= 6x + 4 \in \mathbb{I}_7[x].$$

4.4 Algorithm for reconstructing r and $r(x)$

We now describe an algorithm whose time of execution is $O(n^2)$ (see Borodin and Munro [1975], Lipson [1981]) for computing the solution to a system of n linear congruences over the Euclidean domain ($\mathbb{I}$ or $F[x]$).

4.4.1 *Algorithm*

>**begin**
>>**comment** This algorithm takes $r_i \equiv r \bmod p_i$ $(i = 0, 1, \ldots, n-1)$ and reconstructs r over $\mathbb{I}_M$ or $F[x]$.
>>$M \leftarrow 1;\ R \leftarrow r_0$
>>**for** $k = 1$ **until** $(n-1)$ **do**
>>>**begin**
>>>>$M \leftarrow Mp_{k-1};$
>>>>$\mu \leftarrow M^{-1} \bmod p_k;$
>>>>$d \leftarrow (r_k - R)\mu \bmod p_k;$
>>>>$R \leftarrow R + dM$
>>>**end.**
>>>$R = r;$
>
>**end.**

4.4.2 *Validity of the algorithm.* (i) over $\mathbb{I}_M$. First let us prove the validity for reconstructing r over $\mathbb{I}_M$. Let r be the unique integer $<M$, satisfying the simultaneous congruence relations

$$r_i \equiv r \bmod p_i \qquad (i = 0, 1, \ldots, n-1),$$

where $M = \prod_{i=0}^{n-1} p_i$ and p_i's are pairwise relatively prime. In order to reconstruct r we assume that r is expressed uniquely in the mixed-radix form

$$r = d_0 + d_1 p_0 + d_2 p_0 p_1 + \cdots + d_{n-1} p_0 \cdots p_{n-2},$$

where $d_0, d_1 \cdots d_n$ are standard mixed-radix digits satisfying the inequalities

$$0 \leqslant d_i < p_i \qquad (i = 0, 1, \ldots, n-1).$$

Notice that the primary role played by p_i is to establish a bound on d_i. The algorithm described above essentially determines the digits d_i $(i = 1, 2, \ldots, n-1)$ using r_i and p_i. This is done as follows: Let $M_k = \prod_{i=1}^{k} p_{i-1};\ M_0 = 1$ and

$$\begin{aligned}
R_k &= d_0 + d_1 p_0 + \cdots + d_k p_0 \cdots p_{k-1} \\
&= d_0 + d_1 M_1 + \cdots + d_k M_k.
\end{aligned}$$

It is easily seen that

$$r \bmod p_0 = d_0 = r_0$$
$$r \bmod p_1 = r_1 = d_0 + d_1 p_0;$$

thus $d_1 = (r_1 - d_0) \cdot p_0^{-1} \bmod p_1$. Similarly,

$$d_2 = \frac{r_2 - (d_0 + d_1 p_0)}{p_0 \cdot p_1} \bmod p_2$$

or

$$d_2 = \frac{(r_2 - R_1)}{p_0 \cdot p_1} \bmod p_2$$

and

$$d_k = \frac{(r_k - R_{k-1})}{\displaystyle\prod_{i=1}^{k} p_{i-1}} \cdot \bmod p_k$$

$$= (r_k - R_{k-1}) \cdot M_k^{-1} \cdot \bmod p_k;$$

thus $R_{n-1} = r$.

(ii) over $F[x]$. In the case of reconstruction of $r(x)$ over $\mathbb{I}_p$ we replace r by $r(x)$, and $p_0, p_1, \ldots, p_{n-1}$ respectively by $p_0(x) = (x - x_0)$, $p_{n-1}(x) = (x - x_{n-1})$ and $r(x_i) = r_i$ $(i = 0, 1, \ldots, n-1)$. We also denote $M_k(x) = \prod_{i=1}^{k} p_{i-1}(x)$; $M_0(x) = 1$. We assume that $r(x)$ is an $(n-1)$th degree polynomial expressed uniquely by the Newton's interpolation formula, Blum [1972],

$$r(x) = d_0 + d_1(x - x_0) + \cdots + d_{n-1}(x - x_0) \cdots (x - x_{n-2}),$$

where

$$d_i \in \mathbb{I}_p (= F).$$

The remaining steps to prove the validity are identical to the case of $\mathbb{I}_M$, and we obtain

$$R_{n-1}(x) = r(x).$$

4.4.3 EXAMPLE. (i) Let

$$p_0 = 13, \qquad p_1 = 11, \qquad p_2 = 7$$
$$r_0 = 4, \qquad r_1 = 2, \qquad r_2 = 4.$$

The algorithmic steps are shown below:

k	r_k	p_k	M	μ	d	R_k
0	4	13	1	—	—	4
1	2	11	13	6	10	134
2	4	7	143	5	1	277

Thus $R_2 = r = 277$.

(ii) Let

$$p_0(x) = x - 1, \qquad p_1(x) = x - 2, \qquad p_2(x) = x - 3$$
$$r_0 = r(x_0) = 3, \qquad r_1 = r(x_1) = 2, \qquad r_2 = r(x_2) = 1$$

over $(\mathbb{I}_7, +, \cdot)$.

Then the algorithmic steps are as shown below:

k	r_k	$p_k(x)$	$M(x)$	μ	d	$R_k(x)$
0	3	$(x-1)$	1	—	—	3
1	2	$(x-2)$	$(x-1)$	1	6	$6x+4$
2	1	$(x-3)$	$(x-1)(x-2)$	4	0	$6x+4$

Note: $M^{-1}(x) \bmod p_k(x) = M^{-1}(x) \bmod (x - x_k) = M^{-1}(x_k) \in \mathbb{I}_p$.

4.5 Residue representation of signed integers

We now briefly recall how we represent both positive and negative integers in the residue or modular representation (see Chapter I, Gregory and Krishnamurthy [1984]). Let m be an odd integer. If we form the set

$$\mathbb{S}_m = \left\{ -\frac{m-1}{2}, \ldots, -2, -1, 0, 1, 2, \ldots, \frac{m-1}{2} \right\}$$

then each integer $b \in \mathbb{I}$ can be mapped onto an integer $s \in \mathbb{S}_m$ by the following mapping.

4.5.1 **Definition.** $/\cdot/_m : \mathbb{I} \to \mathbb{S}_m$ is defined by writing

$$/b/_m \equiv s \pmod{m}$$

if and only if $b \equiv s \pmod{m}$ and

$$\frac{-m}{2} < s < \frac{m}{2}.$$

We call $/b/_m$, the symmetric residue of $b \bmod m$.

It is easily verified that $(\mathbb{S}_m, +, \cdot)$ is a finite commutative ring and if m is a prime then $(\mathbb{S}_m, +, \cdot)$ is a finite field. Also $(\mathbb{S}_m, +, \cdot)$ is isomorphic to $(\mathbb{I}_m, +, \cdot)$. Therefore, if the data describing a problem consists of integers in $\mathbb{S}_m$, we can map them into $\mathbb{I}_m$, carry out the computations in $\mathbb{I}_m$ and map the results back into $\mathbb{S}_m$.

The forward mapping function for doing this is

$$|a|_m = \begin{cases} /a/_m & \text{if } 0 \leq |a|_m < \dfrac{m}{2} \\ /a/_m + m, & \text{otherwise} \end{cases}$$

and the inverse mapping function for recovering $s \in \mathbb{S}_m$ is

$$/a/_m = \begin{cases} |a|_m, & \text{if } 0 \leq a_m < \dfrac{m}{2} \\ |a|_m - m, & \text{otherwise.} \end{cases}$$

4.5.2 Remark. It is obvious that the above mappings 4.5.1 can be extended to multiple moduli representation with $M = \prod_{i=0}^{n} p_i$, an odd integer.

4.5.3 Example. Let $m = 101$, then

$$\mathbb{S}_{101} = \{-50, -49, \ldots, 0, \ldots, 50\}$$

$-31 \in \mathbb{I}$ maps to 70.

$+31 \in \mathbb{I}$ maps to 31.

4.5.4 Example. Let $p_0 = 3$, $p_1 = 5$, $p_2 = 7$; $r_0 = 0$, $r_1 = 0$, $r_2 = 4$. Then by using Algorithm 4.4 we obtain $r = 60$. Since $r > 50$, the negative integer corresponding to r is $(60 - 105) = -45$.

5 Polynomials in Several Variables

5.1 Definition. A polynomial $f(x_1, x_2, \ldots, x_n)$ in n variables $x_1, x_2, \ldots, x_n$ over a field F is the sum of a finite number of terms of the form

$$x_1^{\beta_1}, x_2^{\beta_2}, \ldots, x_n^{\beta_n} \quad \text{for all } \beta_i \geq 0$$

with coefficients from F.

5.1.1 *Equality.* Two polynomials $f(x_1, x_2, \ldots, x_n)$ and $g(x_1, x_2, \ldots, x_n)$ are called *equal* if the coefficients of like terms are equal.

5.1.2 *Degree.* If a polynomial $f(x_1, x_2, \ldots, x_n)$ is given over a field F, then its degree with respect to the indeterminate or unknown x_i, $i = 1, 2, \ldots, n$ is the highest exponent with which x_i appears in the terms of the polynomial.

We call the number $\beta_1 + \beta_2 + \cdots + \beta_n$ (that is the sum of the exponents of the unknown) *the degree of the term* $x_1^{\beta_1}, x_2^{\beta_2}, \ldots, x_n^{\beta_n}$. *The degree of the polynomial* is the highest degree of its terms. As in the single variable case we assign the degree zero for constant polynomials and degree $-\infty$ to a zero polynomial.

5.1.3 *Operations.* The operations of addition and multiplication are defined as follows, for the polynomials in n unknowns over a field F.

5.1.4 *Sum.* The sum of the polynomials $f(x_1, x_2, \ldots, x_n)$ and $g(x_1, x_2, \ldots, x_n)$ is a polynomial whose coefficients are obtained by adding the corresponding coefficients of the polynomials f and g.

5.1.5 *Product.* The product of two monomials is defined by the equation

$$ax_1^{\beta_1}x_2^{\beta_2} \cdots x_n^{\beta_n} \cdot bx_1^{\gamma_1}x_2^{\gamma_2} \cdots x_n^{\gamma_n} = (a \cdot b)x_1^{\beta_1+\gamma_1}x_2^{\beta_2+\gamma_2} \cdots x_n^{\beta_n+\gamma_n}$$

after which the product of the polynomials $f(x_1, x_2, \ldots, x_n)$ and $g(x_1, x_2, \ldots, x_n)$ is defined as a termwise multiplication and subsequent collecting and adding of like terms.

5.1.6 *Degree of product.* The degree of a product of two nonzero polynomials in n variables is equal to the sum of the degrees of the polynomials.

5.1.7 *Integral domains.* Given this definition of operations, the collection of polynomials in n unknowns over the field F becomes a commutative ring which does not contain divisors of zero; this ring will be denoted by $F[x_1, x_2, \ldots, x_n]$. Note that $F[x_1, x_2, \ldots, x_n]$ is an integral domain (Section 2.3).

5.1.8 *Field of rational functions.* Since $F[x_1, x_2, \ldots, x_n]$ is an integral domain, it has a field of quotients; we denote this by $F(x_1, \ldots, x_n)$.

$$F(x_1, \ldots, x_n) = \left\{ \frac{a(x_1, \ldots, x_n)}{b(x_1, \ldots, x_n)} : a, b \in F[x_1, \ldots, x_n], b \neq 0 \right\}$$

5.1.9 *Recursive representation.* In the recursive representation, we rewrite a polynomial in several variables $x_1, \ldots, x_n$ as a single variable polynomial, by choosing one of the variables, say x_1, as the main variable, with its coefficients lying in the ring of polynomials in the auxiliary variables $x_2, \ldots, x_n$.
Thus

$$a(x_1, \ldots, x_n) = \sum_{i=0}^{\deg(x_1)} a_i(x_2, \ldots, x_n) x_1^i$$

where $\deg(x_1)$ denotes the highest degree of x_1 in $a(x_1, \ldots, x_n)$. Such a representation is useful for computational purposes as will be seen later in Sections 5.3 and 5.4.

5.2 Formal power series

By analogy with the single variable case (Section 3.1), we can define a formal power series in n variables; (Chisholm [1973, 1974, 1977], Baker and Grave-Morris [1981], Baker [1975]). For simplicity, we illustrate this using a two-variable example. Consider the rational polynomial in two variables x, y over a field F

(5.1)
$$\frac{a(x,y)}{b(x,y)} = \alpha(x,y) = \frac{\sum_{i,j=0}^{m} a_{ij} x^i y^i}{\sum_{i,j=0}^{m} b_{ij} x^i y^j} \in F(x, y),$$

where $a_{ij}, b_{ij} \in F = (\mathbb{Q}_p, +, \cdot)$. We assume that $a_{00} \neq 0$, $b_{00} \neq 0$.
We can expand $b(x,y)$ as a formal power series over F provided we

define the formal reciprocal of $b(x,y)$ by

(5.2)
$$\left[\sum_{i,j=0}^{m} b_{ij}x^iy^j\right] \cdot \left[\sum_{i,j=0}^{\infty} d_{ij}x^iy^j\right] = 1$$

so that

(5.3)
$$\left[\sum_{i,j=0}^{m} b_{ij}x^iy^j\right]^{-1} = \sum_{i,j=0}^{\infty} d_{ij}x^iy^j.$$

In (5.2), the unity element in F is denoted by 1. We now truncate

$$\left[\sum_{i,j=0}^{m} a_{ij}x^iy^j\right] \cdot \left[\sum_{i,j=0}^{\infty} b_{ij}x^iy^j\right]^{-1}$$

to a required number of terms, so that the degree of the terms in the product is less than or equal to $(r-1)$. This means

(5.4)
$$c(x,y) = a(x,y)b^{-1}(x,y) = \sum_{i,j=0} c_{ij}x^iy^j + O(r)$$

where $O(r)$ denotes those terms of the form x^ky^{r-k} with degree greater than or equal to r. We denote this truncation by $c(x,y)$ $O(r)$.

5.2.1 EXAMPLE. Let
$$\frac{a(x,y)}{b(x,y)} = \frac{1+2x+y}{1+x+2y},$$

where $a_{ij}, b_{ij} \in (\mathbb{I}_7, +, \cdot)$ then

$$c(x,y) = (1+2x+y)(1+x+2y)^{-1}$$
$$= (1+2x+y)(1+6x+5y+x^2+4y^2+4xy+6x^3+6y^3$$
$$+ x^2y+2xy^2+O(4))$$

or

$$c(x,y)\,O(4) = 1+x+6y+6x^2+2y^2+6xy+x^3+3y^3+3x^2y.$$

5.2.2 EXAMPLE. Let
$$\frac{a(x,y)}{b(x,y)} = \frac{3+5y}{1+5x+5y}$$

where $a_{ij}, b_{ij} \in (\mathbb{I}_7, +, \cdot)$

$$c(x,y)\,O(4) = 3x^3+2y^3+x^2y+5x^2+y^2+6xy+6x+4y+3.$$

5.3 Pseudodivision and remainder theorem

We earlier defined Euclidean domains in Section 2.5. Two important examples of integral domains which are not Euclidean domains are the ring $\mathbb{I}[x]$ of polynomials in a variable x over the ring of integers $\mathbb{I}$, and the ring $R[x,y]$ of polynomials in variables x and y over the real numbers R. These two rings are therefore not principal ideal rings. Therefore while dealing with polynomials in several variables over a field F, the conventional Euclidean algorithm is not directly applicable. However, such a polynomial can be rewritten in a recursive canonical form (Section 5.1.9) as a single variable polynomial, by choosing one of the variables, say x_1, as the main variable with its coefficients lying in the ring of polynomials in the auxiliary variables $x_2, \ldots, x_n$. This is helpful to define a pseudodivision process, resembling the Euclidean division, for polynomials in several variables.

5.3.1 EXAMPLE. Consider the polynomials

$$a(x) = 13x^4 + 11x^3 + 7x^2 + 5x + 3 \in \mathbb{I}[x]$$
$$b(x) = 19x^2 + 17x + 13 \in \mathbb{I}[x].$$

Here the Euclidean division process with $a(x)$ as the dividend and $b(x)$ as the divisor cannot be performed, since the coefficients of the quotient $q(x)$ do not belong to $\mathbb{I}$.

However, by performing a simple trick, it is possible to carry out an operation resembling the division (hence called pseudodivision). This trick consists in multiplying $a(x)$ by a suitable power i of the leading coefficient of $b(x)$, where $i = \deg a(x) - \deg b(x) + 1$, so that $q(x) \in \mathbb{I}[x]$. Thus $19^3 \cdot a(x) = q(x) \cdot b(x) + r(x)$.

5.3.2 *Pseudodivision.* In general, if $a(x) = a_m x^m + \cdots + a_0$ and $b(x) = b_n x^n + \cdots + b_0$, $b_n \neq 0$ and $m \geq n$, then the pseudodivision is defined by $b_n^{m-n+1} \cdot a(x) = q(x) \cdot b(x) + r(x)$ such that $\deg r(x) < \deg b(x)$, and $q(x) \in \mathbb{I}[x]$. We call $q(x)$ as pseudoquotient and $r(x)$ as pseudoremainder.

5.3.3 *Pseudodivision of polynomials in several variables.* The above notion of pseudodivision can be extended to polynomials in several variables $x_1, \ldots, x_n$ thus:

Let $\deg(x_1)$ $[a(x_1, \ldots, x_n)]$ and $\deg(x_1)$ $[b(x_1, \ldots, x_n)]$ denote the highest degree of x_1 in $a(x_1, \ldots, x_n)$ and $b(x_1, \ldots, x_n)$, respectively. Let $d = \deg(x_1)$ $[a(x_1, \ldots, x_n)] - \deg(x_1)$ $[b(x_1, \ldots, x_n)] \geq 0$ and let $\beta(x_2, \ldots, x_n)$ denote the coefficient of the highest power of x_1 in $b(x_1, \ldots, x_n)$. Then the pseudodivision operation of $a(x_1, \ldots, x_n)$ by $b(x_1, \ldots, x_n)$ results in

(5.5) $\quad \beta^{d+1} a(x_1, \ldots, x_n) = q(x_1, \ldots, x_n) b(x_1, \ldots, x_n) + r(x_1, \ldots, x_n).$

5.3.4 *Pseudoremainder theorem.* If we let $b = x_1 - c$ where $c \in F$, then we have from (5.5)

(5.6) $\qquad\qquad\qquad a(x_1, \ldots, x_n) = q(x_1 - c) + r.$

Thus for $x_1 = c \in F$

(5.7) $\qquad\qquad\qquad a(c, \ldots, x_n) = r(c, x_2, \ldots, x_n)$

which is a generalization of the remainder Theorem 3.5.1. Equation (5.7) corresponds to the evaluation of $a(x_1, \ldots, x_n)$ at $x_1 = c$, and is therefore a polynomial in $x_2, \ldots, x_n$. We can therefore use the concept of residue or remainder as in Section 4.3 to describe the evaluation of a polynomial as below.

$$(5.8) \qquad a(x_1, \ldots, x_n) \bmod (x_1 - c) = r(c, \ldots, x_n).$$

The evaluation (5.8) can be made for each variable assuming some value over F. The question now arises as to whether it is possible to reconstruct the polynomial $a(x_1, \ldots, x_n)$ from the evaluated values. The answer is "yes" and we can generalize the Lagrangian interpolation (Section 4.3) for this purpose, provided we have the adequate number of evaluations that would enable us to determine the coefficients of the polynomial $a(x_1, \ldots, x_n)$ uniquely. For example, if each variable $x_i (i = 1, 2, \ldots, n)$ has a maximum degree $\leq d$ then we need to have $(d+1)^n$ points of evaluation.

In the next section we will explain the procedure for the interpolation of polynomials in several variables using the extended Lagrangian interpolation.

5.4 Evaluation–interpolation

The extension of Lagrangian interpolation to polynomial in several variables can be carried out in a manner analogous to the single variable interpolation, if we use the recursive representation (Section 5.1.9).

Let $r(x,y)$ be any two variable polynomial of maximum degree m in x and n in y, and let $r(x,y) \in F[x,y]$. The problem of interpolation consists in reconstructing $r(x,y)$ given its evaluated values at $(m+1) \cdot (n+1)$ points, $r_{ij} = r(x_i, y_j)$, $i = 0, 1, \ldots, m$; $j = 0, 1, \ldots, n$. Algorithm 4.4 can be used twice for carrying out this interpolation: once for interpolating $r(x_i, y) = r_i(y)$ for each x_i $(i = 0, 1, \ldots, m)$, using the values $r(x_i, y_j)$ $(j = 0, 1, \ldots, n)$; at the second time we take $r_i(y)$, which parametrically defines the residues for each x_i, as inputs to reconstruct $r(x,y)$.

It is clear that such an algorithm when extended recursively to n-variables, would treat functions and arguments to be of equal status, and a program variable assumes values drawn from a set of functions (values defined parametrically) rather than numerical values. Present day development in applicative languages and related computer architecture permits us to use such mixed data types during a computation, (Turner [1979], Kluge [1983]).

5.4.1 EXAMPLE. Let $r(x,y) \in \mathbb{I}_7[x,y]$ be given by the following table; it is assumed that x and y have both a maximum degree ≤ 2.

x	y	r(x,y)
0	0	2
0	1	6
0	2	5
1	0	6
1	1	4
1	2	4
2	0	5
2	1	4
2	2	5

We first use $r(0,0) = 2$, $r(0,1) = 6$, and $r(0,2) = 5$ in Algorithm 4.4 to obtain the following table:

k	r_k	$p_k(y)$	$M(y)$	μ	d	$R_k(y)$
0	2	y	1	—	—	2
1	6	$y-1$	y	1	4	$4y+2$
2	5	$y-2$	$y(y-1)$	4	1	y^2+3y+2

Thus $r(0,y) = y^2 + 3y + 2$.

In a similar way, we find $r(1,y)$; and $r(2,y)$.

k	r_k	$p_k(y)$	$M(y)$	μ	d	$R_k(y)$
0	6	y	1	—	—	6
1	4	$y-1$	y	1	5	$5y+6$
2	4	$y-2$	$y(y-1)$	4	1	y^2+4y+6

Thus $r(1,y) = y^2 + 4y + 6$.

k	r_k	$p_k(y)$	$M(y)$	μ	d	$R_k(y)$
0	5	y	1	—	—	5
1	4	$y-1$	y	1	6	$6y+5$
2	5	$y-2$	$y(y-1)$	4	1	y^2+5y+5

Thus $r(2,y) = y^2 + 5y + 5$.

We now use $r(0,y)$, $r(1,y)$, and $r(2,y)$ as the new values of residues in

Algorithm 4.4 and obtain the table:

k	r_k	$p_k(x)$	$M(x)$	μ	d	$R_k(x,y)$
0	y^2+3y+2	x	1	—	—	y^2+3y+2
1	y^2+4y+6	$x-1$	x	1	$y+4$	$y^2+xy+4x+3y+2$
2	y^2+5y+5	$x-2$	$x(x-1)$	4	1	$x^2+y^2+xy+3x+3y+2$

Thus $r(x,y) = x^2+y^2+xy+3x+3y+2$.

5.4.2 Remark. In the general case of interpolation of n-variable polynomials in which each variable occurs with a degree $\leq d$, we use the evaluated values of $r(x_1,\ldots,x_n)$ at the $(d+1)^n$ points $x_{1k}, x_{2k}, \ldots, x_{nk}$ for each $k = 1, 2, \ldots, (d+1)$. We then use, for a given k, the values $\{x_{1k}, x_{2k}, \ldots, x_{(n-1)k}, x_{nj}\}$, $j = 1, 2, \ldots, (d+1)$ and obtain the parametric residues $r(x_{1k}, x_{2k}, \ldots, x_{(n-1)k}, x_n)$ in x_n. We then use these k parametric residues and reiterate the algorithm 4.4 for each one variable in succession, until we totally reconstruct $r(x_1, \ldots, x_n)$. In numerical analysis, such an approximation or interpolation is known as product operator method (see Schumaker [1977], McKinney [1972], Hartley [1976]). We will see more about this in the next chapter.

EXERCISES I

1. Find the gcd(601,101).

2. Find the gcd(3125,195).

3. Find the lcm(25,15).

4. Find the mod 125 inverse of 17.

5. Find the mod 19 inverse of 11.

6. Find the mod 601 inverse of 101.

7. Find the mod 97 inverse of 14.

8. Find $\left|\frac{11}{17}\right|_{625}$.

9. Find $\left|\frac{11}{17}\right|_{601}$.

10. Find $/{-279}/_{1001}$ with $p_0 = 7$, $p_1 = 11$, $p_2 = 13$.

11. Divide $11x^4+8x^3+7x^2+5x+9$ by $12x^2+9x+7$ in $\mathbb{I}_{13}[x]$.

12. Divide $17x^4+13x^3+11x^2+7x+5$ by $19x^3+8x^2+x+17$ in $\mathbb{I}_{101}[x]$.

13. Find the inverse transform of $a(x) = 2x^2+3x+4 \in \mathbb{I}_{11}[x]$ modulo x^6.

14. Find the inverse transform of $a(x) = 1+300x+2x^2 \in \mathbb{I}_{601}[x]$ up to the sixth degree term.

15. Find the power series representation (up to fourth degree term) of

$$\frac{2x^2+5x+1}{6x^2+6x+1} \in F_{601}(x).$$

16. Find the power series representation (up to second degree) of

$$\frac{x+1}{2x+1} \in F_{17}(x).$$

17. Evaluate the polynomial

$$11x^4+8x^3+9x^2+2x+7 \in \mathbb{I}_{19}[x]$$

at $x = 1, 2, 3, 4, 5$. Reconstruct the polynomial using Lagrangian interpolation.

18. Evaluate the polynomial

$$100x^4+8x^3+7x^2+6x+5 \in \mathbb{I}_{101}[x]$$

at $x = 10, 20, 30, 40, 50$. Reconstruct using the Lagrangian interpolation.

19. Evaluate the two-variable polynomial

$$r(x,y) = 2xy + 3x + 3y - 3 \in \mathbb{I}_{11}[x,y]$$

at $(x_i, y_i) = (0,0), (0,1), (1,0), (1,1)$ and reconstruct using Lagrangian interpolation.

20. Evaluate the polynomial

$$x^2+y^2+z^2+3xy+3yx+3zx \in \mathbb{I}_{101}[x,y,z]$$

at 27 distinct points in $\mathbb{I}_{101}$ and reconstruct.

21. Evaluate the polynomial

$$5x^3+6x^2+7x+8 \in \mathbb{I}[x]$$

at $x = 1, 2, 3, 4$ using multiple modulus residue arithmetic with three primes $p_0 = 7$, $p_1 = 11$, $p_2 = 13$ and combine the results by the Chinese remainder theorem, to obtain the evaluated values in $\mathbb{I}$.

22. Find the $\gcd(3x^2+5x+7, 2x^2+9x+8)$ over $\mathbb{I}_{17}[x]$.

23. Find the continued fraction expansion of 11/7.

24. Find the continued fraction expansion of $(x^2+3)/(x+1)$ over $F_7(x)$.

25. Find the continued fraction expansion of $x^3/(1+16x+2x^2)$ over $F_{17}(x)$.

CHAPTER II

Polynomial Matrix—Evaluation, Interpolation, Inversion

1 Introduction

In many problems in systems theory, (Barnett [1971], Rosenbrock and Storey [1970]) we encounter matrices (called "Polynomial matrices") whose elements are polynomials over the field of rationals or over the ring of integers, in an indeterminate x or several indeterminates $x, y, z, \ldots$. The inversion and manipulation of such matrices become involved due to the following reasons:

(i) The conventional floating point arithmetic is not suitable for this purpose, since we cannot neglect very small terms occurring as the coefficients (elements from a field or a ring) of a polynomial.

(ii) The integer arithmetic demands excessively long precision operands and the computation becomes slow.

(iii) The coefficients of the polynomial grow during the inversion (or other transformation operations) since Gaussian type reduction does not allow one to use the division operation when the coefficients of the polynomial matrix elements are integers.

Accordingly, it is necessary to use procedures which are computationally more economical and at the same time give exact results.

In this chapter, we describe a procedure known as the evaluation–interpolation method which is suitable for exact inversion of polynomial matrices. This method consists in evaluating the given polynomial matrix at certain specified points (elements of a field or a ring) and obtaining a set of numerical matrices with elements over the field of rationals or the ring of integers. These matrices are then inverted exactly (Gregory and Krishnamurthy [1984]) in the usual manner. Since every one of the corresponding elements of these inverted matrices correspond to the evaluated values of the inverse (when it exists) of the polynomial matrix at these specified points, we can use the Lagrange or other related interpolation schemes to reconstruct the inverse of the polynomial matrix.

Here we shall first consider the case of matrices whose elements are polynomials in a single indeterminate (single-variable polynomial) and later extend the method to matrices whose elements are polynomials in several indeterminates (multivariable polynomials).

The interpolation–evaluation schemes in this chapter will be described with respect to polynomial matrices (with integer coefficients in elements), where the evaluation is carried out at distinct integral values. We will explain in the next chapter how it is more economical to use Fourier methods for this purpose.

Before we proceed any further we will summarize the relevant results from matrix theory.

2 Results from Matrix Theory

2.1 Matrices over fields

Let A be an $m \times n$ matrix whose elements a_{ij} ($1 \le i \le m$, $1 \le j \le n$) belong to a field F. We now define the following elementary transformations on A.

2.1.1 *Elementary transformations.* There are three types of elementary row or column transformations that can be performed on A.

2.1.2 *Type 1.* Let $i \ne j$ and B be the matrix obtained from A by interchanging its ith row (column) and jth row (column). Then B is said to be obtained from A by an elementary row (column) transformation of type 1.

Note that the type 1 row (column) transformation is equivalent to multiplying A on left (right) by the permuted (ith row and jth row interchanged) identity matrix, known as the permutation matrix $P = T_1$

$$
T_1 = \begin{array}{c} \\ i \rightarrow \\ \\ j \rightarrow \\ \\ \\ \end{array}
\begin{bmatrix}
1 & \text{-} & \text{-} & \text{-} & \text{-} & \text{-} & \text{-} & 0 \\
 & \ddots & & & & & \\
 & & 0 & \text{-} & \text{-} & 1 & \\
 & & \vdots & & \vdots & & \\
 & & 1 & \text{-} & \text{-} & 0 & \\
0 & \text{-} & \text{-} & \text{-} & \text{-} & \text{-} & \text{-} & 1
\end{bmatrix}
$$

In other words

$$\text{Type 1 row:} \qquad T_1 A = B$$
$$\text{Type 1 column:} \quad A T_1 = B.$$

2.1.3 *Type 2.* Let $c \neq 0 \in F$ and B be the matrix obtained by the addition to the ith row of A, c times its jth row. Then B is said to be obtained from A by an elementary row transformation of type 2. Note that the type 2 row transformation is equivalent to multiplying A on the left by the matrix T_{2R}.

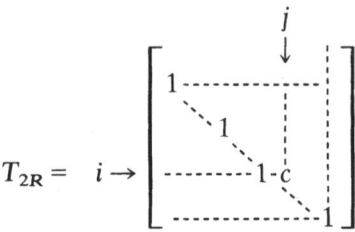

In other words $T_{2R} A = B$.

The same operation for column is equivalent to multiplication of A on the right by T_{2C}.

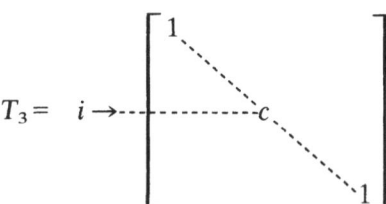

In other words $A T_{2C} = B$.

2.1.4 *Type 3.* Let $c \neq 0 \in F$ and B be the matrix obtained as the result of the multiplication of the ith row (column) of A by c. Then B is said to be obtained from A by an elementary transformation of type 3. Note that the type 3 row (column) transformation is equivalent to multiplying A on the left (right) by T_3.

$$T_3 = \quad i \rightarrow \begin{bmatrix} 1 & & & & \\ & \ddots & & & \\ & & c & & \\ & & & \ddots & \\ & & & & 1 \end{bmatrix}$$

2.1.5 *Equivalence.* Let A and B be two $m \times n$ matrices over F and let B be obtained from A by the successive application of finitely many arbitrary elementary transformations. Then we say that A is equivalent to B in F and denote $A \cong B$.

Note that the property of equivalence is an equivalence relation:

(i)   If $A \cong B$ and $B \cong C$ then $A \cong C$;
(ii) if $A \cong A$;
(iii) if $A \cong B$ implies $B \cong A$.

Also in type 2 transformation, if $c = 0$, and in type 3 transformation, if $c = 1$, then these are identity transformations.

2.1.6 *Determinants.* Let B be a square matrix. The symbol

$$D = \begin{vmatrix} b_{11} \cdots b_{1k} \\ b_{k1} \cdots b_{kk} \end{vmatrix}$$

stands for a k-rowed determinant of order k. It is defined as the sum of $k!$ terms of the form

$$(-1)^i b_{1i_1} b_{2i_2} \cdots b_{ki_k},$$

where the sequence of subscripts $i_1, \ldots, i_k$ ranges over all permutations of $1, 2, \ldots, k$ and the permutation of $i_1, \ldots, i_k$ may be carried into $1, 2, \ldots, k$ by interchanges. The sign $(-1)^i$ is unique. We denote determinant of B by $\det B$. If B is obtained by type 1 elementary transformation from A then $\det B = -\det A$; if B is obtained by type 2 elementary transformation from A then $\det B = \det A$; if B is obtained by type 3 elementary transformation from A then $\det B = c \cdot \det A$.

2.1.7 **Theorem.** *Every nonzero $m \times n$ matrix A of rank r is equivalent to a matrix*

$$U = \begin{bmatrix} I_r & 0 \\ 0 & 0 \end{bmatrix} \quad over\ F,$$

where I_r is the $r \times r$ identity matrix. (See Chapter III, Gregory and Krishnamurthy [1984].)

2.1.8 **Remark.** Thus from Theorem 2.1.7 we can write

$$PAQ = U,$$

where P and Q are nonsingular square matrices (which are products of the elementary transformation matrices T_1, T_{2R}, T_{2C}, T_3). When A is a square nonsingular matrix ($\det A \neq 0$ over F) then $A^{-1} = QP$, since $U = I$ and $\det A = (\det P \cdot \det Q)^{-1}$.

2.1.9 **Remark.** In this chapter, we will confine our discussions mainly to nonsingular matrices. In Chapters IV and V we shall deal with singular matrices and generalized inverses over principal ideal rings and integral domains.

2.2 Matrices over $F[x]$

Let F be a field, we designate by $F[x]$ the set of all polynomials in x with coefficients in F. We shall consider an m by n matrix $A(x)$ with elements in $F[x]$. It is clear that such a matrix can also be treated as a polynomial

$$A(x) = A_0 + A_1 x + \cdots + A_k x^k$$

in which the coefficients A_0, $A_1, \ldots, A_k$ are $m \times n$ matrices whose elements belong to F. We can therefore call $A(x)$ either as a polynomial matrix or a matrix polynomial of order $(m \times n)$. When A_i's are not matrices but merely elements over F, then the matrix polynomial reduces to a scalar polynomial.

2.2.1 *Degree.* The degree of a matrix polynomial is the largest integer k for which $A_k \neq 0$; the coefficient A_k of a polynomial $A(x)$ of degree k is therefore called the leading coefficient. As in the scalar polynomial, the degree of zero matrix polynomial is undefined.

2.2.2 *Equality.* Two matrix polynomials $A(x)$ and $B(x)$ of order $(m \times n)$ given by

$$A(x) = A_0 + A_1 x + \cdots + A_k x^k$$
$$B(x) = B_0 + B_1 x + \cdots + B_k x^k$$

are equal if $A_i = B_i$ $(i = 0, 1, \ldots, k)$.

2.3 Elementary transformations

We can define three types of elementary transformations on $A(x)$ as on A (Section 2.1) over F. However, we have some restrictions.

In type 2, the quantities c are permitted to be any element in $F[x]$. However, those of type 3 are restricted so that the quantity $c \in F[x]$ shall have an inverse in $F[x]$. Then c must necessarily be a constant polynomial $\neq 0$, or any nonzero element (unit) of F.

2.3.1 *Equivalence.* Two $(m \times n)$ polynomial matrices $A(x)$ and $B(x)$ are equivalent in $F[x]$, if there exists a sequence of elementary transformations carrying A into B.

In other words, we call two matrices A and B equivalent in $F[x]$ if there exists square matrices P and Q such that

$$PAQ = B.$$

Since P and Q are products of elementary transformation matrices, each having a determinant 1 or a constant c, the determinants of P and Q are also units or nonzero constants over F.

The field $F(x)$ of all rational functions of x with coefficients in F contains $F[x]$ and so if A and B are equivalent in $F[x]$, they are also equivalent in $F(x)$. Obviously, A and B are equivalent in $F[x]$ only if they have the same rank.

2.4 Smith Canonical form

In Section 2, Chapter IV, Gregory and Krishnamurthy [1984] we introduced the notion of Smith Canonical form for a matrix A defined over the ring of integers I. As already mentioned in Chapter I, since both I and $F[x]$ are Euclidean domains we can define the Smith Canonical form for A over $F[x]$.

2.4.1 **Theorem.** *Every nonzero matrix A of rank r with elements in $F[x]$ is equivalent in $F[x]$ to*

$$\begin{bmatrix} S & 0 \\ 0 & 0 \end{bmatrix},$$

where $S = \mathrm{diag}\{s_{11}(x), s_{22}(x), \ldots, s_{rr}(x)\}$ for monic polynomials $s_{ii}(x)$ such that $s_{ii}(x)$ divides $s_{i+1,i+1}(x)$ for $1 \le i \le r-1$. For proof, see Gantmacher [1959], Albert [1956].

2.4.2 **Remark.** In fact, for Euclidean domains Theorem 2.4.1 is generalized as below.

2.4.3 **Theorem.** *Every $m \times n$ matrix A of rank r can be written as*

$$P \begin{bmatrix} S & 0 \\ 0 & 0 \end{bmatrix} Q,$$

where P and Q are square matrices with a constant nonzero determinant. $S = \mathrm{diag}\{s_{11}, s_{22}, \ldots, s_{rr}\}$ and s_{ii} form the complete set of nonassociates. Also s_{ii} divides $s_{i+1,i+1}$ for $1 \le i \le r-1$ and the product $\prod_{i=1}^{r} s_{ii}$ is the greatest common divisor of all the $r \times r$ subdeterminants of A.

The form S is unique and is called the Smith Canonical form or Smith normal form. For proof, see Newman [1972], MacDuffee [1948].

2.4.4 *Invariant factors.* The polynomials $s_{ii}(x)$ are called the invariant polynomials of the matrix A because they are unaltered by equivalence transformations; Barnett [1971], Rosenbrock and Storey [1970].

2.4.5 **Remark.** If A is an integer matrix the determinants of P and Q in Theorem 2.4.3 is ± 1 (since these are the only units in I).

2.4.6 EXAMPLE. Let

$$A = \begin{bmatrix} x+1 & 2x+3 \\ x+2 & 3x+4 \end{bmatrix}$$

and

$$B = \begin{bmatrix} 2x+4 & x+1 \\ 4x+5 & 6x+1 \end{bmatrix} \quad \text{over } \mathbb{I}_7[x],$$

then

$$A+B = \begin{bmatrix} 3x+5 & 3x+4 \\ 5x & 2x+5 \end{bmatrix}$$

$$A \cdot B = \begin{bmatrix} 3x^2+5 & 6x^2+x+4 \\ 4x & 5x^2+2x+6 \end{bmatrix}.$$

2.5 Substitution

The operation of substitution of $c \in F$ for the indeterminate x in $A(x)$ results in the evaluation of each element $a_{ij}(x)$ for $x = c$. We denote this operation by $A(x) \bmod(x-c)$ to mean that if

$$A(x) = (a_{ij}(x)), \quad \text{then } A(c) = (a_{ij}(c)) \equiv (a_{ij} \bmod(x-c)) \equiv A(x) \bmod(x-c).$$

We will now state the following elementary theorems without proof; in all these theorems the matrices A, B, C, D are over $F[x]$.

2.5.1 **Theorem.** *The following statements are equivalent*:

(i) $A \equiv B \bmod(x-x_i), \qquad x_i \in F$;
(ii) $B \equiv A \bmod(x-x_i), \qquad x_i \in F$;
(iii) $A - B \equiv 0 \bmod(x-x_i), \qquad x_i \in F$.

2.5.2 **Theorem.** *If $A \equiv B \bmod(x-x_i)$ and $B \equiv C \bmod(x-x_i)$ then $A \equiv C \bmod(x-x_i)$ where $x_i \in F$.*

2.5.3 **Theorem.** *If A and B are conformable then*

$$(A \pm B) \bmod(x-x_i) = A \bmod(x-x_i) \pm B \bmod(x-x_i).$$

2.5.4 **Theorem.** *If $A \equiv B \bmod(x-x_i)$ and $C \equiv D \bmod(x-x_i)$ then*

$$\alpha A + \beta C \equiv \alpha B + \beta D \bmod(x-x_i), \quad \text{where } \alpha, \beta \in F.$$

2.5.5 **Theorem.** *If A and B are conformable then*

$$A \cdot B \bmod(x-x_i) \equiv A \bmod(x-x_i) \cdot B \bmod(x-x_i).$$

2.5.6 **Theorem.** *If A and C and B and D are conformable and if*

$$A \equiv B \bmod(x-x_i)$$
$$C \equiv D \bmod(x-x_i)$$

then $AC \equiv BD \bmod(x-x_i)$.

2.6 Inverse matrix

For square matrices $A(x)$, $C(x) \in F(x)$, if $A(x_i) \cdot C(x_i) = C(x_i) \cdot A(x_i) = I$ (where $x_i \in F$ and I is the identity matrix) then we write $C(x_i) = A^{-1}(x_i)$ or $C(x) \bmod(x - x_i) \equiv A^{-1}(x) \bmod(x - x_i)$, and call $C(x_i)$ the multiplicative inverse of $A(x_i)$ over F.

2.6.1 **Definition.** $A(x)$ is said to be nonsingular modulo $(x - x_i)$, if and only if, both $\det A(x) \neq 0$ and $\gcd(\det A(x), (x - x_i)) = 1$; otherwise, $A(x_i)$ is called singular over F.

2.6.2 **Theorem.** $A^{-1}(x_i)$ *exists over F if and only if $A(x_i)$ is nonsingular over F.*

2.6.3 **Theorem.** *If $A^{-1}(x) \bmod(x - x_i)$ exists then it is unique.*

PROOF. Assume C and D are both multiplicative inverses of $A(x_i)$; then $CA = AC = I$ and $DA = AD = I$. Therefore

$$C \bmod(x - x_i) = CI \bmod(x - x_i)$$
$$= C(AD) \bmod(x - x_i) = CA(D) \bmod(x - x_i)$$
$$= D \bmod(x - x_i). \qquad \square$$

2.6.4 *Minor, cofactor.* A minor $M_{ij}(x)$ of an $n \times n$ matrix $A(x)$ is the $(n-1) \times (n-1)$ matrix obtained from A by striking out row i and column j. The cofactor A_{ij} of a_{ij} is defined as

$$A_{ij}(x) = (-1)^{i+j} \det M_{ij}(x).$$

2.6.5 *Adjoint.* The adjoint of A is defined by $\operatorname{adj} A = (A_{ji})$, or in other words the ijth element of $\operatorname{adj} A$ is A_{ji}.

2.6.6 *Inverse.* The inverse of A is related to the adjoint A by

$$(\operatorname{adj} A)A = A(\operatorname{adj} A) = I \cdot \det A,$$

where I is the identity matrix.

2.6.7 **Remark.** All the Theorems 2.5.1–2.6.6 can be extended to matrices $A(x_1, x_2, \ldots, x_n)$ by virtue of the recursive representation 5.1.9 and the pseudoremainder theorem 5.3.4 of Chapter I.

2.6.8 *Degree of $A^{-1}(x)$.* The maximal degree of $A^{-1}(x)$ (see Section 2.2.1) can be estimated using the definition in Section 2.6.6 and assuming $A(x)$ is defined over $F(x)$, the field of rational functions over F. (We, however, assume that $a_{ij}(x)$ have all unit denominators to begin with.) Let $(d-1)$ be the maximal degree of $a_{ij}(x)$. Then $\det A(x)$ can be at most of degree $n(d-1)$ since it is a sum of products of n elements (Section 2.1.6).

Thus the numerator and denominator polynomials in (adj A/det A) can have at most degree $D = n \ (d-1)$.

Therefore, if we use algorithms which interpolate the denominator and numerator separately, we need to evaluate the polynomial matrix at $D+1 = n \ (d-1)+1$ points, at least, so that these can be reconstructed uniquely over $F(x)$. For example, if $n = 10$, $d = 10$, we need to evaluate $A(x)$ at 91 points at least.

2.6.9 *Coefficient size in $A^{-1}(x)$.* When the given matrix $A(x)$ is defined over $\mathbb{I}_p[x]$ or $F_p(x)$, the coefficient size is bounded by the field element $(p-1)$.

However, if $A(x)$ is defined over $\mathbb{Q}(x)$ and to start with, we assume that $a_{ij}(x) \in \mathbb{I}[x]$, then adj A has the same property. Consequently, we can get an estimate of the coefficient size in adj A and det A, which respectively correspond to the numerator and denominator of the inverse of A.

Consider the product of two polynomials each of degree $(d-1)$ with maximal coefficient $m \in \mathbb{I}$, then their product can have a maximal coefficient $m^2 d$. Using the same argument, when n such polynomials are multiplied together, the maximum coefficient size is $m^n \cdot d^{(n-1)}$. Since there are $n!$ such product terms added together to compute det A (Section 2.1.6), the maximal coefficient size of the determinant polynomial has an estimate $E \leqslant n! \ m^n d^{n-1}$.

Naturally, this estimate grows exponentially with n and is one of the essential difficulties in the polynomial matrix computations; see Kannan [1983], Chou and Collins [1982], Frumkin [1975]; attempts to prove better bounds have so far not been successful.

Therefore, in order to make the computations feasible one should use multiple moduli residue arithmetic with several primes or the finite segment p-adic arithmetic, Gregory and Krishnamurthy [1984], as described in Section 3.2 below.

3 Matrix Method—Evaluation and Interpolation of Single Variable Polynomials

We now describe the basic matrix method for the evaluation and interpolation technique for polynomials in single variables. Later we shall extend these techniques to the multivariable polynomials in Section 4.

Let us denote by $\mathscr{P}_{ij}(x)$ the polynomial at the ijth entry in A. We assume that $\mathscr{P}_{ij}(x)$ is of degree $(d-1)$. Then we can write

(3.1)
$$\mathscr{P}_{ij}(x) = \sum_{i=0}^{d-1} p_i x^i.$$

We can rewrite this expression (3.1) as an inner product thus:

$$(3.2) \qquad \mathcal{P}_{ij}(x) = [1 \ x \cdots x^{d-1}] \begin{bmatrix} p_0 \\ p_1 \\ \vdots \\ p_{d-1} \end{bmatrix} = X^T P,$$

where

$$P^T = [p_0 p_1 \cdots p_{d-1}]$$
$$X^T = [1 \ x \cdots x^{d-1}].$$

Thus if in (3.2) we replace x by the values x_i $(i = 0, 1, 2 \cdots)$, $\mathcal{P}_{ij}(x_i)$ can be evaluated.

Equation (3.2) corresponds to the *synthesis* (passage from coefficients to functional values) or the building up of the function. On the other hand, if the $\mathcal{P}_{ij}(x_i)$ are given, then the vectors P_{ij} (coefficients of $\mathcal{P}_{ij}(x)$) are computed using

$$(3.3) \qquad P_{ij} = (X_i^T)^{-1} \mathcal{P}_{ij}(x_i),$$

where

$$X_i^T = \begin{bmatrix} 1 & x_0 & x_0^2 & \cdots & x_0^{d-1} \\ 1 & x_1 & x_1^2 & \cdots & x_1^{d-1} \\ \vdots & \vdots & \vdots & \cdots & \vdots \\ 1 & x_{d-1} & x_{d-1}^2 & \cdots & x_{d-1}^{d-1} \end{bmatrix}.$$

Equation (3.3) corresponds to *analysis* (passage from functional values to the coefficients) or interpolation. Also note that X_i^T is usually ill-conditioned.

3.1 Single variable polynomial matrix inversion

Let $A(x)$ be the given polynomial matrix where the coefficients of $\mathcal{P}_{ij}(x)$ are integers and let us assume that $A(x)$ and $A^{-1}(x)$ are defined over the quotient field $\mathbb{Q}(x)$. The estimates of D (maximal degree in $A^{-1}(x)$) and E (maximal coefficient in $A^{-1}(x)$) are computed using the method described in Sections 2.6.8 and 2.6.9.

The method of evaluation–interpolation for the inversion of a nonsingular $A(x)$ consists of the following steps.

3.1.1 *Algorithm.*

Comment: No limit is set on the size of the coefficients of the polynomial. Since X_i^T is ill-conditioned, we assume that exact inverse of X_i^T is used in solving (3.3).

Step 1: Evaluate $A(x)$ at $(D+1)$ distinct points, x_i $(i = 0, 1, \ldots, D)$ using (3.1).

Step 2:   Invert each $A(x_i)$ using an exact method and compute its determinant (Gregory and Krishnamurthy [1984]; see also Section 2.1.8 of this chapter).

Comment: If there were unlucky choices of x_i, $\det A(x_i)$ may be zero. Then choose a new x_i, reevaluate $A(x_i)$ and compute $A^{-1}(x_i)$; repeat this process until we have $(D+1)$ distinct points for which $A^{-1}(x_i)$ exist.

Step 3:   Interpolate each one of the corresponding entries and the determinant using (3.3).

3.1.2 EXAMPLE. Let the given polynomial matrix $A(x)$ be given by

$$A(x) = \begin{bmatrix} x & x+1 & 0 \\ x+2 & 0 & -x+1 \\ 0 & x+1 & x+2 \end{bmatrix}.$$

Let $x_0 = 0$, $x_1 = 1$, $x_2 = 2$, $x_3 = 3$. We evaluate the polynomials at 4 distinct points since the maximum degree D of the determinant is estimated to be 3; the coefficient size $E \leqslant 192$.

We have now the matrices $A(x_0)$, $A(x_1)$, $A(x_2)$ and $A(x_3)$ with determinants and adj obtained by any exact method described in Gregory and Krishnamurthy [1984].

$$A(x_0) = \begin{bmatrix} 0 & 1 & 0 \\ 2 & 0 & 1 \\ 0 & 1 & 2 \end{bmatrix}; \qquad A^{-1}(x_0) = -\frac{1}{4}\begin{bmatrix} -1 & -2 & 1 \\ -4 & 0 & 0 \\ 2 & 0 & -2 \end{bmatrix}$$

$$A(x_1) = \begin{bmatrix} 1 & 2 & 0 \\ 3 & 0 & 0 \\ 0 & 2 & 3 \end{bmatrix}; \qquad A^{-1}(x_1) = -\frac{1}{18}\begin{bmatrix} 0 & -6 & 0 \\ -9 & 3 & 0 \\ 6 & -2 & -6 \end{bmatrix}$$

$$A(x_2) = \begin{bmatrix} 2 & 3 & 0 \\ 4 & 0 & -1 \\ 0 & 3 & 4 \end{bmatrix}; \qquad A^{-1}(x_2) = -\frac{1}{42}\begin{bmatrix} 3 & -12 & -3 \\ -16 & 8 & 2 \\ 12 & -6 & -12 \end{bmatrix}$$

$$A(x_3) = \begin{bmatrix} 3 & 4 & 0 \\ 5 & 0 & -2 \\ 0 & 4 & 5 \end{bmatrix}; \qquad A^{-1}(x_3) = -\frac{1}{76}\begin{bmatrix} 8 & -20 & -8 \\ -25 & 15 & 6 \\ 20 & -12 & -20 \end{bmatrix}.$$

We now interpolate each entry in adj A and the determinant using (3.3). Here we have

$$X_i^T = \begin{bmatrix} 1 & 0 & 0 & 0 \\ 1 & 1 & 1 & 1 \\ 1 & 2 & 4 & 8 \\ 1 & 3 & 9 & 27 \end{bmatrix}$$

and

$$(X_i^T)^{-1} = \begin{bmatrix} 1 & 0 & 0 & 0 \\ \dfrac{-11}{6} & 3 & \dfrac{-3}{2} & \dfrac{1}{3} \\ 1 & \dfrac{-5}{2} & 2 & \dfrac{-1}{2} \\ \dfrac{-1}{6} & \dfrac{1}{2} & \dfrac{-1}{2} & \dfrac{1}{6} \end{bmatrix}$$

Thus, for example, the determinant polynomial has coefficients given by (3.3)

$$(X_i^T)^{-1} \begin{bmatrix} -4 \\ -18 \\ -42 \\ -76 \end{bmatrix} = \begin{bmatrix} -4 \\ -9 \\ -5 \\ 0 \end{bmatrix}$$

or the determinant polynomial is $-(4+9x+5x^2)$. Repeating this procedure for each entry we obtain

$$A^{-1}(x) = \frac{1}{-(4+9x+5x^2)} \begin{bmatrix} -1+x^2 & -2-3x-x^2 & 1-x^2 \\ -4-4x-x^2 & 2x+x^2 & x^2-x \\ 2+3x+x^2 & -x-x^2 & -2-3x-x^2 \end{bmatrix}.$$

3.2 Modular method

If the entries of $A(x_i)$ are very large integers, it is computationally more advantageous to use multiple modulus (or p-adic) arithmetic for evaluating $A(x_{ij}) \bmod p_j$ for each $x_i \in \mathbb{I}_{p_j}$, compute $A^{-1}(x_{ij}) \bmod p_j$ for each i, j, and use the Chinese remainder theorem or related procedures (Section 4.4, Chapter I) to reconstruct $A^{-1}(x)$. The choice of number of points, as before depends on D and so we evaluate $A(x)$ for $x_{ij} \bmod p_j$ ($i = 0, 1, \ldots, D$) with $D = n\,(d-1)$. The choice of primes is thus decided (see Section 2.6.9). Let $M = \prod_{j=0}^{k} p_j$; then $-M/2 < E < M/2$ or $M > 2E = 2[n!\,m^n d^{n-1}]$, where n is order of the matrix A, m the maximal size of coefficient in $A_{ij}(x)$, and $(d-1)$ the maximal degree of any element $a_{ij}(x)$.

3.2.1 *Algorithm.* The algorithm is essentially the same as in 3.1.1 with some minor changes:

In *Step 1*, the evaluation is carried out for each prime p_j.
In *Step 2*, each $A(x_{ij})$ is inverted mod p_j.
In *Step 3*, each $A(x_{ij})$ is interpolated mod p_j to obtain $A^{-1}(x) \bmod p_j$; and finally combined using the Chinese remainder theorem to obtain $A^{-1}(x) \bmod M$.

3.2.2 **Remarks.** It is possible that at Step 2, the det $A(x_{ij})$ becomes zero modulo p_i; then this unlucky choice is to be eliminated and a new x_{ij} or new p_i is to be chosen.

For interpolatory Step 3, we can either use Chinese remaindering based on Lagrangian interpolation (Algorithm 4.4 of Chapter I) or the matrix method (3.3) directly over the field $(\mathbb{I}_{p_i}, +, \cdot)$. The coefficients mod $\mathbb{I}_{p_i}$ are then combined using the Chinese remainder theorem to get $A^{-1}(x)$ mod M.

3.2.3 EXAMPLE. We now illustrate the modular method for the Example 3.1.2. For this example, we have $m = 2$, $n = 3$, $d = 2$; $E = 3! \, 2^3 \cdot 2^2 = 192$, and $D = 3$. We therefore choose three primes 5, 11 and 13 (Note that the choice of $p_j = 7$ for $x_{ij} = 2$, results in the singularity of $A(2)$ mod 7); thus $M = 715$.

(i) $p_0 = 5$: $x_0 = 0$; $x_1 = 1$; $x_2 = 2$; $x_3 = 3$.

$$A(x_0) = \begin{bmatrix} 0 & 1 & 0 \\ 2 & 0 & 1 \\ 0 & 1 & 2 \end{bmatrix}; \qquad A^{-1}(x_0) = 1 \begin{bmatrix} 4 & 3 & 1 \\ 1 & 0 & 0 \\ 2 & 0 & 3 \end{bmatrix}$$

$$A(x_1) = \begin{bmatrix} 1 & 2 & 0 \\ 3 & 0 & 0 \\ 0 & 2 & 3 \end{bmatrix}; \qquad A^{-1}(x_1) = \frac{1}{2} \begin{bmatrix} 0 & 4 & 0 \\ 1 & 3 & 0 \\ 1 & 3 & 4 \end{bmatrix}$$

$$A(x_2) = \begin{bmatrix} 2 & 3 & 0 \\ 4 & 0 & 4 \\ 0 & 3 & 4 \end{bmatrix}; \qquad A^{-1}(x_2) = \frac{1}{3} \begin{bmatrix} 3 & 3 & 2 \\ 4 & 3 & 2 \\ 2 & 4 & 3 \end{bmatrix}$$

$$A(x_3) = \begin{bmatrix} 3 & 4 & 0 \\ 0 & 0 & 3 \\ 0 & 4 & 0 \end{bmatrix}; \qquad A^{-1}(x_3) = \frac{1}{4} \begin{bmatrix} 3 & 0 & 2 \\ 0 & 0 & 1 \\ 0 & 3 & 0 \end{bmatrix}.$$

(ii) $p_1 = 11$: $x_0 = 0$; $x_1 = 1$; $x_2 = 2$; $x_3 = 3$.

$$A(x_0) = \begin{bmatrix} 0 & 1 & 0 \\ 2 & 0 & 1 \\ 0 & 1 & 2 \end{bmatrix}; \qquad A^{-1}(x_0) = \frac{1}{7} \begin{bmatrix} 10 & 9 & 1 \\ 7 & 0 & 0 \\ 2 & 0 & 9 \end{bmatrix}$$

$$A(x_1) = \begin{bmatrix} 1 & 2 & 0 \\ 3 & 0 & 0 \\ 0 & 2 & 3 \end{bmatrix}; \qquad A^{-1}(x_1) = \frac{1}{4} \begin{bmatrix} 0 & 5 & 0 \\ 2 & 3 & 0 \\ 6 & 9 & 5 \end{bmatrix}$$

$$A(x_2) = \begin{bmatrix} 2 & 3 & 0 \\ 4 & 0 & 10 \\ 0 & 3 & 4 \end{bmatrix}; \qquad A^{-1}(x_2) = \frac{1}{2} \begin{bmatrix} 3 & 10 & 8 \\ 6 & 8 & 2 \\ 1 & 5 & 10 \end{bmatrix}$$

$$A(x_3) = \begin{bmatrix} 3 & 4 & 0 \\ 5 & 0 & 9 \\ 0 & 4 & 5 \end{bmatrix}; \qquad A^{-1}(x_3) = 1 \begin{bmatrix} 8 & 2 & 3 \\ 8 & 4 & 6 \\ 9 & 10 & 2 \end{bmatrix}.$$

(iii) $p_2 = 13$: $x_0 = 0$; $x_1 = 1$; $x_2 = 2$; $x_3 = 3$.

$$A(x_0) = \begin{bmatrix} 0 & 1 & 0 \\ 2 & 0 & 1 \\ 0 & 1 & 2 \end{bmatrix}; \quad A^{-1}(x_0) = \frac{1}{9}\begin{bmatrix} 12 & 11 & 1 \\ 9 & 0 & 0 \\ 2 & 0 & 11 \end{bmatrix}$$

$$A(x_1) = \begin{bmatrix} 1 & 2 & 0 \\ 3 & 0 & 0 \\ 0 & 2 & 3 \end{bmatrix}; \quad A^{-1}(x_1) = \frac{1}{8}\begin{bmatrix} 0 & 7 & 0 \\ 4 & 3 & 0 \\ 6 & 11 & 7 \end{bmatrix}$$

$$A(x_2) = \begin{bmatrix} 2 & 3 & 0 \\ 4 & 0 & 12 \\ 0 & 3 & 4 \end{bmatrix}; \quad A^{-1}(x_2) = \frac{1}{10}\begin{bmatrix} 3 & 1 & 10 \\ 10 & 8 & 2 \\ 12 & 7 & 1 \end{bmatrix}$$

$$A(x_3) = \begin{bmatrix} 3 & 4 & 0 \\ 5 & 0 & 11 \\ 0 & 4 & 5 \end{bmatrix}; \quad A^{-1}(x_3) = \frac{1}{2}\begin{bmatrix} 8 & 6 & 5 \\ 1 & 2 & 6 \\ 7 & 1 & 6 \end{bmatrix}.$$

Determinants: On interpolation for each p_i:

$$\det A(x) \bmod 5 = 1 + x$$
$$\det A(x) \bmod 11 = 7 + 2x + 6x^2$$
$$\det A(x) \bmod 13 = 9 + 4x + 8x^2.$$

Combining coefficients using the Chinese remainder theorem (Section 4.4, Chapter I):

$$\det A(x) \bmod 715 = 711 + 706x + 710x^2$$

which corresponds to $\det A(x) = -4 - 9x - 5x^2$. The other entries are similarly combined and we obtain:

$$A^{-1}(x) = \frac{1}{-(4 + 9x + 5x^2)}\begin{bmatrix} -1+x^2 & -2-3x-x^2 & 1-x^2 \\ -4-4x-x^2 & 2x+x^2 & x^2-x \\ 2+3x+x^2 & -x-x^2 & -2-3x-x^2 \end{bmatrix}.$$

3.2.4 **Remark.** For the exact solution of systems of linear equations with polynomial coefficients and rational function coefficients, see McClellan [1973], [1977].

4 Tensor Product Method—Evaluation and Interpolation of Multivariable Polynomials

Multivariable polynomial matrices often occur in the area of multivariable realization theory and multidimensional filtering (Bose [1983]). The matrix computations involving multivariable polynomials get complicated due to the fact that these polynomials are not Euclidean domains (and hence unique quotient and remainder are not obtained in the division) and they are only integral domains (and hence only the cancellation law is valid). In fact one of the major computational difficulties is the coefficient growth problem which is not controllable, since division is not allowed.

For inverting the multivariable polynomial matrices, we can use the evaluation–interpolation technique, as we did for the single variable case.

The evaluation–interpolation technique which we are going to discuss, also becomes computationally involved due to:

(i)   The large number of evaluations that need to be carried out for each one of the m variables $x_1, x_2, \ldots, x_m$ at each point of the multi-dimensional grid.
(ii) The inversion of a large number of numerical matrices.
(iii) The back interpolation of a large number of polynomials.

A more economical method is to do the evaluation and interpolation using the Fourier methods. We will discuss this in the next chapter.

In this section we will briefly introduce the notion of tensor (or Kronecker) product evaluation–interpolation scheme over the multi-dimensional grid, in order to invert a multivariable polynomial matrix. Although numerous other methods can be used for this purpose, the tensor product method is economical and more convenient to use (Gordon [1971], Schumaker [1977], Delvos and Pasdorf [1977], Pereyra and Scherer [1973]).

We first begin with evaluation—interpolation technique in a two-dimensional grid and later extend this to higher dimensions using the tensor product method.

4.1 Two-dimensional grid evaluation–interpolation

Let us first assume that the given matrix has elements which are two-variable polynomials. The most general two-variable polynomial is of the type:

$$(4.1) \qquad \mathscr{P}(x, y) = \sum_{i=0}^{n-1} \sum_{j=0}^{n-1} p_{ij} x^i x^j.$$

We can rewrite (4.1) in the bilinear form

$$(4.2) \qquad \mathscr{P}(x,y) = [1\ y\ y^2 \cdots y^{n-1}] \begin{bmatrix} p_{00} & p_{10} \cdots p_{n-1,0} \\ & \vdots \\ & \vdots \\ p_{0,n-1} \cdots p_{n-1,n-1} \end{bmatrix} \begin{bmatrix} 1 \\ x \\ x^2 \\ \vdots \\ x^{n-1} \end{bmatrix}$$

$$= Y^T P^T X,$$

say, where

$$Y^T = [1 \; y \cdots y^{n-1}]$$
$$X^T = [1 \; x \cdots x^{n-1}]$$
$$P^T = (p_{ij}) = \text{coefficients of the polynomial.}$$

To evaluate (4.2) for known p_{ij}, we need only replace the Y^T and X by their values. For n point evaluation, for each variable x and y, we will have to use the points (x_i, y_j) in the two-dimensional grid,

$$0 \le i \le n-1, \qquad 0 \le j \le n-1.$$

If we assume that $x_i = 0, 1, 2, \ldots, (n-1)$ and $y_j = 0, 1, 2, \ldots, (n-1)$ are the points on the grid, then the evaluated values are:

$$\mathcal{P}(x_i, y_j) = \begin{bmatrix} 1 & 0 & \cdots & 0 \\ 1 & 1 & \cdots & 1 \\ 1 & 2 & \cdots & 2^{n-1} \\ \cdot & \cdot & \cdots & \cdot \\ 1 & (n-1) & \cdots & (n-1)^{n-1} \end{bmatrix} P^T \begin{bmatrix} 1 & 1 & \cdots & 1 \\ 0 & 1 & 2 \cdot & (n-1) \\ 0 & \cdot & \cdots & \cdots \\ \cdot & \cdot & \cdots & \cdots \\ 0 & 1 & 2^{(n-1)} & (n-1)^{n-1} \end{bmatrix}$$
(4.3)

$$= Y_j^T P^T X_i.$$

Note Y_j^T and X_i are Vandermonde matrices and hence nonsingular; however, they are usually highly ill-conditioned; therefore, we need to use exact methods for inversion.

4.1.1 EXAMPLE. Let $\mathcal{P}(x, y) = -3 + 3x + 3y + 2xy$ over $\mathbb{I}$.
If we choose $x_0 = 0$, $x_1 = 1$, $y_0 = 0$, $y_1 = 1$, we obtain

$$\mathcal{P}(x_i, y_j) = \begin{bmatrix} 1 & y_0 \\ 1 & y_1 \end{bmatrix} \begin{bmatrix} p_{00} & p_{10} \\ p_{01} & p_{11} \end{bmatrix} \begin{bmatrix} 1 & 1 \\ x_0 & x_1 \end{bmatrix}$$

$$= \begin{bmatrix} 1 & 0 \\ 1 & 1 \end{bmatrix} \begin{bmatrix} -3 & 3 \\ 3 & 2 \end{bmatrix} \begin{bmatrix} 1 & 1 \\ 0 & 1 \end{bmatrix} = \begin{bmatrix} -3 & 0 \\ 0 & 5 \end{bmatrix}.$$

Also, given $\mathcal{P}(x_i, y_j)$ the computation of P^T is done from (4.3) using

(4.4) $(Y_j^T)^{-1} \mathcal{P}(x_i, y_j) X_i^{-1} = P^T.$

4.1.2 EXAMPLE. Given

$$\mathcal{P}(x_i, y_j) = \begin{bmatrix} 5 & 10 \\ 10 & 17 \end{bmatrix}$$

at $x_0 = 1$, $x_1 = 2$; $y_0 = 1$, $y_1 = 2$; find P.

Using (4.4)

$$P^T = \begin{bmatrix} 1 & y_0 \\ 1 & y_1 \end{bmatrix}^{-1} \begin{bmatrix} 5 & 10 \\ 10 & 17 \end{bmatrix} \begin{bmatrix} 1 & 1 \\ x_0 & x_1 \end{bmatrix}^{-1}$$

$$= \begin{bmatrix} 1 & 1 \\ 1 & 2 \end{bmatrix}^{-1} \begin{bmatrix} 5 & 10 \\ 10 & 17 \end{bmatrix} \begin{bmatrix} 1 & 1 \\ 1 & 2 \end{bmatrix}^{-1}$$

$$= \begin{bmatrix} 2 & -1 \\ -1 & 1 \end{bmatrix} \begin{bmatrix} 5 & 10 \\ 10 & 17 \end{bmatrix} \begin{bmatrix} 2 & -1 \\ -1 & 1 \end{bmatrix}$$

$$= \begin{bmatrix} -3 & 3 \\ 3 & 2 \end{bmatrix}$$

$$= \begin{bmatrix} p_{00} & p_{10} \\ p_{01} & p_{11} \end{bmatrix}.$$

Thus $\mathscr{P}(x,y) = -3 + 3x + 3y + 2xy$.

4.2 Tensor products and their properties

In order to simplify our notation and also to algorithmize the evaluation–interpolation procedure, we now need to introduce the concept of Kronecker or Tensor product (Rao and Mitra [1971], Davis [1979], Graham [1982]).

Let $A(m \times n) = (a_{ij})$ and $B(p \times q) = (b_{ij})$ be two matrices. Then the Kronecker product (denoted by $\otimes$) is defined as

$$A \otimes B = (a_{ij}B),$$

and is an $(mp \times nq)$ matrix with $a_{ij}B$ as its ijth partition; or

$$A \otimes B = \begin{bmatrix} a_{11}B & a_{12}B & \cdots & a_{1n}B \\ \vdots & \vdots & & \vdots \\ a_{m1}B & a_{m2}B & \cdots & a_{mn}B \end{bmatrix}.$$

The following are some of the important properties of the tensor product of matrices:

(i) $\alpha \cdot A \otimes \beta \cdot B = \alpha \cdot \beta \cdot (A \otimes B);$ $\alpha, \beta \in F$ or $\mathbb{R}$.
(ii) $(A + B) \otimes C = A \otimes C + B \otimes C.$
(iii) $A \otimes (B + C) = A \otimes B + A \otimes C$ (+ denotes conformable matrix addition).
(iv) $A \otimes (B \otimes C) = (A \otimes B) \otimes C.$
(v) $(A \otimes B) \cdot (C \otimes D) = (A \cdot C) \otimes (B \cdot D)$ ($\cdot$ denotes conformable matrix multiplication).
(vi) $(A \otimes B)^T = A^T \otimes B^T$ (T denotes transpose).
(vii) $\text{rank}(A \otimes B) = \text{rank } A \cdot \text{rank } B.$

For square $A(m \times m)$, $B(n \times n)$:

(viii) trace$(A \otimes B) = $ trace $A \cdot$ trace B.

(ix) $\det(A \otimes B) = (\det A)^n \cdot (\det B)^m$.

(x)   There exists a permutation matrix P depending on m, n such that $B \otimes A = P^T (A \otimes B) P$.

(xi)   If A and B are nonsingular then $(A \otimes B)^{-1} = A^{-1} \otimes B^{-1}$.

(xii) If A and B have Moore–Penrose inverses A^+ and B^+, then $(A \otimes B)^+ = A^+ \otimes B^+$.

(xiii) If $ACB = D$ is a matrix equation, where A is an $(m \times n)$ matrix, C is an $(n \times r)$ matrix, B is an $(r \times p)$ matrix, and D is an $(m \times p)$ matrix, then it can be rewritten as a matrix-vector equation

$$(A \otimes B^T) v(C) = v(D),$$

where the operation $v(M)$ for any $(m \times n)$ matrix M is the transpose of the row vector obtained by concatenating all the rows of M end to end with the first row on the left and the last row on the right to form a $1 \times mn$ vector.

In fact $v(M)$ results in

$$m_{ij} = [v(M)]_{n(i-1)+j} = v_k,$$

where k denotes the component index of vector v_k.

The above properties are useful for the evaluation–interpolation of multivariable polynomials.

4.3 Use of Tensor product for grid evaluation–interpolation

Consider equation (4.3). Using the property (xiii) of Section 4.2 we can write (4.3) as

(4.6) $$v(\mathscr{P}(x_i, y_j)) = (Y_j^T \otimes X_i^T) v(P^T).$$

Similarly, (4.4) can be rewritten as

(4.7) $$v(P^T) = [(Y_j^T)^{-1} \otimes (X_i^T)^{-1}] v(\mathscr{P}(x_i, y_j)).$$

The equations (4.6) and (4.7) are very useful for extension to many variables. For instance, if we have a three variable polynomial in x, y, z we can generalize them to

(4.8) $$v(\mathscr{P}(x_i, y_j, z_k)) = (Z_k^T \otimes Y_j^T \otimes X_i^T) v(P^T)$$

and

(4.9) $$[v(P^T) = [(Z_k^T)^{-1} \otimes (Y_j^T)^{-1} \otimes (X_i^T)^{-1}] v(\mathscr{P}(x_i, y_j, z_k)).$$

Equation (4.8) yields evaluation while (4.9) yields the interpolation of $\mathscr{P}(x, y, z)$.

In (4.9) if Z_k^T, Y_j^T, and X_i^T are singular, we can use their Moore–Penrose inverses, to obtain the least-square interpolation of $\mathcal{P}(x,y,z)$

$$(4.10) \qquad v(P^T) = [(Z_k^T)^+ \otimes (Y_j^T)^+ \otimes (X_i^T)^+] v(\mathcal{P}(x_i, y_j, z_k))$$

(see Campbell [1982], Campbell and Meyer [1979]).

4.3.1. EXAMPLE. Evaluate

$$\mathcal{P}(x,y,z) = 1 + x + 2x^2 + 2y + xy - x^2y + 2z + zx - zx^2$$
$$+ 3zy - xyz + 2x^2yz$$

at $x_i = 0, 1, 2$; $y_j = 0, 1$; and $z_k = 0, 1, 2$ over the real field. Thus

$$X_i^T = \begin{bmatrix} 1 & 0 & 0 \\ 1 & 1 & 1 \\ 1 & 2 & 4 \end{bmatrix}; \qquad Y_j^T = \begin{bmatrix} 1 & 0 \\ 1 & 1 \end{bmatrix}; \qquad Z_k^T = \begin{bmatrix} 1 & 0 \\ 1 & 1 \\ 1 & 2 \end{bmatrix}.$$

Using (4.8) we obtain the following values for $\mathcal{P}(x_i, y_j, z_k)$. (Table 4.1.) Also given $\mathcal{P}(x_i, y_j, z_k)$ we can compute P^T using (4.10).

Table 4.1 $\mathcal{P}(x_i, y_j, z_k)$

z_k	y_j	x_i	$\mathcal{P}(x_i, y_j, z_k)$
0	0	0	1
0	0	1	4
0	0	2	11
0	1	0	3
0	1	1	6
0	1	2	11
1	0	0	3
1	0	1	6
1	0	2	11
1	1	0	8
1	1	1	12
1	1	2	20
2	0	0	5
2	0	1	8
2	0	2	11
2	1	0	13
2	1	1	18
2	1	2	29

For this example

$$(X^{-1})^T = (X^+)^T = \frac{1}{4} \begin{bmatrix} 4 & 0 & 0 \\ -6 & 8 & -2 \\ 2 & -4 & 2 \end{bmatrix}$$

$$(Y^{-1})^T = (Y^+)^T = \begin{bmatrix} 1 & 0 \\ -1 & 1 \end{bmatrix}$$

$$(Z^+)^T = \frac{1}{6}\begin{bmatrix} 5 & 2 & -1 \\ -3 & 0 & 3 \end{bmatrix}.$$

Thus we can reconstruct P^T using (4.10).

4.3.2 **Remark.** The tensor product method of evaluation and interpolation is also called "Product operator method" due to the fact that we break up the total interpolation process into m independent parts corresponding to each variable $x_1, \ldots, x_m$ and use the tensor product space as the basis.

For instance, if a two-variable function is to be interpolated, we proceed in two stages: interpolation of the first variable; then the interpolation of the second variable using the functional form in the first variable. This stepwise reconstruction is analogous to the Chinese remainder theorem method described in Section 4 of Chapter I.

4.4 Inversion of multivariable polynomial matrices

In order to invert a multivariable polynomial matrix A over the quotient field $\mathbb{Q}(x_1, \ldots, x_m)$, we need to estimate the degree of the determinant and also the size of the coefficients of the polynomial entries of the inverse matrix. For this purpose we assume, as in the single variable case, that the coefficients of the polynomial entries of A are integers.

4.4.1 *Degree estimate.* Let us assume that among the entries a_{ij} of the matrix A, each variable has a maximal degree $(R-1)$ and there are m variables. Since the determinant of an $(n \times n)$ matrix is a sum of product of n elements, the degree of det A $(x_1, \ldots, x_m)$ can be at most of degree $n(R-1)$ in each variable.

Therefore we need to evaluate the matrix A at $[n(R-1)+1]^m$ points for uniquely interpolating the determinant. For example, if $n = 10$, $R = 6$, $m = 3$ we need to evaluate at 132651 points!

4.4.2 *Coefficient estimate.* Let N be the maximal coefficient size among the entries a_{ij}, $(R-1)$ be the maximal degree, and m be the number of variables.

The maximal coefficient size in a product of two such m-variable polynomials with a maximal degree $(R-1)$ in each variable, is $O(R^m N^2)$. When n such polynomials are multiplied, the maximal coefficient size is $O(2^m \cdot R^{m(n-1)} \cdot N^n)$. Since there are $n!$ such terms in a determinant the maximal coefficient estimate is $E = O(n! \cdot 2^m R^{m(n-1)} \cdot N^n)$.

In order to use multiple modulus arithmetic, we need to choose primes

p_i such that

$$M = \prod_i p_i,$$

where $-M/2 < E < M/2$. Thus $M = O(2E)$; the complexity is clearly exponential in nature:

4.4.3 EXAMPLE. Let

$$A = \begin{bmatrix} x+y & x-y & x+y \\ x-y & x+y & x-y \\ x+y & x-y & x^2+y^2 \end{bmatrix} \quad \text{over } \mathbb{Q}(x,y).$$

Since $\det(A)$ can have a degree of 4 in both x and y, we evaluate A at the 25 points (x_i, y_j), where $x_i = \{1, 2, 3, 4, 5\}$; $y_j = \{1, 2, 3, 4, 5\}$. Of these 25 matrices, the sixteen nonsingular matrices corresponding to $(1, 2)$, $(1, 3)$, $(1, 4)$, $(1, 5)$, $(2, 1)$, $(2, 2)$, $(2, 3)$, $(2, 4)$, $(3, 1)$, $(3, 2)$, $(3, 3)$, $(3, 4)$, $(4, 1)$, $(4, 2)$, $(4, 3)$, $(4, 4)$ are used for the interpolation over the rational field, using exact computation.

The interpolation results in

$$\det A = 4x^3 y + 4xy^3 - 4x^2 y - 4xy^2$$

$$A^{-1} = \begin{bmatrix} a_{11} & a_{12} & a_{13} \\ a_{21} & a_{22} & a_{23} \\ a_{31} & a_{32} & a_{33} \end{bmatrix},$$

where

$$a_{11} = x^3 + xy^2 + x^2 y + y^3 - x^2 - y^2 + 2xy$$
$$a_{12} = x^2 - y^2 + y^3 - x^3 + x^2 y - xy^2$$
$$a_{13} = -4xy$$
$$a_{21} = x^2 - y^2 - x^3 + y^3 + x^2 y - xy^2$$
$$a_{22} = x^3 + y^3 + xy^2 + x^2 y - 2xy - x^2 - y^2$$
$$a_{23} = 0$$
$$a_{31} = -4xy$$
$$a_{32} = 0$$
$$a_{33} = 4xy.$$

EXERCISES II

1. Using elementary transformations find the rank, determinant, and inverse of

$$\begin{bmatrix} 2 & 3 & 4 \\ 3 & 4 & 5 \\ 4 & 5 & 6 \end{bmatrix} \quad \text{over } (\mathbb{I}_7, +, \cdot).$$

2. Find the determinant and inverse of

$$\begin{bmatrix} 8 & 5 & 7 \\ 9 & 6 & 8 \\ 5 & 8 & 3 \end{bmatrix} \quad \text{over } (\mathbb{I}_{13}, +, \cdot).$$

3. Find the determinant of

$$\begin{bmatrix} 75 & 80 & 15 \\ 23 & 17 & 18 \\ 19 & 21 & 32 \end{bmatrix} \quad \text{over } \mathbb{I}$$

using multiple modulus arithmetic, and Chinese remainder theorem.

4. Multiply the following two matrices over $\mathbb{I}$ using multiple modulus arithmetic and Chinese remainder theorem. Verify your result.

$$\begin{bmatrix} 7 & 8 & 9 \\ 5 & 6 & 8 \\ 15 & 11 & 8 \end{bmatrix} \cdot \begin{bmatrix} 11 & 18 & 16 \\ 13 & 11 & 19 \\ 11 & 17 & 14 \end{bmatrix}$$

5. Multiply the following two matrix polynomials over integers using evaluation and interpolation.

$$\begin{bmatrix} 1+x & 3+x \\ 5+x & 6+x \end{bmatrix} \cdot \begin{bmatrix} 2+x & 5+x \\ 9+x & 8+x \end{bmatrix}$$

6. Multiply the following two matrix polynomials over $\mathbb{I}_{101}[x]$.

$$\begin{bmatrix} 15x+19 & 17x+18 \\ 16x+14 & 19x+21 \end{bmatrix} \cdot \begin{bmatrix} 90x+17 & 100x+1 \\ 80x+19 & x+100 \end{bmatrix}$$

7. In the problem 6, use evaluation–interpolation over $\mathbb{I}_{101}$ and carry out the same operation; explain how you decide on the number of points for evaluation.

8. Invert the matrix

$$\begin{bmatrix} 8 & 7 \\ 7 & 8 \end{bmatrix} \otimes \begin{bmatrix} 9 & 5 \\ 5 & 9 \end{bmatrix}$$

over $(\mathbb{I}_{37}, +, \cdot)$.

9. Find the Kronecker product of

$$\begin{bmatrix} 17 & 18 & 19 \\ 20 & 17 & 21 \\ 23 & 16 & 8 \end{bmatrix} \otimes \begin{bmatrix} 3 & 6 \\ 9 & 11 \end{bmatrix}$$

over $(\mathbb{I}_{53}, +, \cdot)$.

10. Find the characteristic equation of

$$\begin{bmatrix} 15-\lambda & 25 \\ 29 & 17-\lambda \end{bmatrix}$$

over $(\mathbb{I}_{23}, +, \cdot)$.

11. Simplify the following expression of matrices over $(\mathbb{I}_{17}, +, \cdot)$.

$$\left[\begin{pmatrix} 5 & 3 \\ 7 & 9 \end{pmatrix} \otimes \begin{pmatrix} 8 & 2 \\ 7 & 6 \end{pmatrix} - \begin{pmatrix} 7 & 1 \\ 9 & 6 \end{pmatrix} \otimes \begin{pmatrix} 4 & 3 \\ 2 & 1 \end{pmatrix} \right] \cdot \begin{bmatrix} 3 & 4 \\ 5 & 6 \end{bmatrix}$$

12. Find the Kronecker product of

$$\begin{pmatrix} 1 & x \\ x^2 & x \end{pmatrix} \otimes \begin{pmatrix} 2 & x \\ x & x^2 \end{pmatrix}$$

over $(\mathbb{I}_{17}, +, \cdot)$.

13. Using evaluation–interpolation find the product of

$$\begin{pmatrix} 1 & x \\ 1 & y \end{pmatrix} \cdot \begin{pmatrix} x & y \\ y & x \end{pmatrix}$$

over the real field.

14. Find

$$\begin{bmatrix} 1 & x \\ 1 & x^2 \end{bmatrix}^3$$

over $(\mathbb{I}_{41}, +, \cdot)$ using evaluation–interpolation method.

15. Evaluate the following polynomial matrix A at $x = 0, 1, 2, 3$ and use it to find the inverse of A over the quotient field $\mathbb{Q}(x)$

$$A = \begin{bmatrix} x & x+1 \\ 2x & 3x \end{bmatrix}.$$

16. Evaluate the following polynomial matrix A at $x = 0, 1, 2$, $y = 0, 1, 2$, and find the inverse of A over the quotient field $\mathbb{Q}(x,y)$

$$A = \begin{bmatrix} x & y \\ x+y & y^2 \end{bmatrix}.$$

17. Find the inverse of the matrix A over the quotient field $\mathbb{Q}(x, y, z)$

$$A = \begin{bmatrix} x & y & z \\ y & z & x \\ z & x & y \end{bmatrix}$$

using the evaluation–interpolation method.

18. Find the inverse of the following matrix over $F_3(x,y)$

$$\begin{bmatrix} 2x & y+2 \\ x+1 & y+1 \end{bmatrix}$$

using evaluation–interpolation method.

19. Invert the matrix $A \otimes B = C$ where

$$A = \begin{bmatrix} 1 & y \\ y^2 & y \end{bmatrix}; \quad B = \begin{bmatrix} 1 & x \\ x & x^3 \end{bmatrix}$$

over the real field.

20. Find the characteristic equation $\det(A - \lambda I)$ using the evaluation–interpolation method:

$$(A - \lambda I) = \begin{bmatrix} 1-\lambda & 2 & 3 \\ 3 & 4-\lambda & 5 \\ 5 & 6 & 7-\lambda \end{bmatrix};$$

where the elements of A are real.

21. Solve for $C_{11}, C_{12}, C_{21}, C_{22}$:

$$\begin{bmatrix} 3 & 4 \\ 7 & 10 \end{bmatrix} \begin{bmatrix} C_{11} & C_{12} \\ C_{21} & C_{22} \end{bmatrix} \begin{bmatrix} 8 & 9 \\ 6 & 5 \end{bmatrix} = \begin{bmatrix} 13 & 13 \\ 0 & 0 \end{bmatrix}$$

over $(\mathbb{I}_{17}, +, \cdot)$ using the property (xiii) of Kronecker product (Section 4.2).

22. Solve for $C_{11}, C_{12}, C_{21}, C_{22}$:

$$\begin{bmatrix} 5 & 8 \\ 9 & 8 \end{bmatrix} \begin{bmatrix} C_{11} & C_{12} \\ C_{21} & C_{22} \end{bmatrix} \begin{bmatrix} 4 & 3 \\ 3 & 8 \end{bmatrix} = \begin{bmatrix} 3 & 0 \\ 9 & 0 \end{bmatrix}$$

over $(\mathbb{I}_{11}, +, \cdot)$ using the property (xiii) of Kronecker product (Section 4.2).

23. Find the rank of

$$\begin{bmatrix} 1 & 2 & 3 \\ 4 & 3 & 2 \\ 2 & 3 & 4 \end{bmatrix} \otimes \begin{bmatrix} 1 & 2 \\ 3 & 4 \end{bmatrix}$$

over $(\mathbb{I}_5, +, \cdot)$.

24. Find the determinant of

$$\begin{bmatrix} 1 & x \\ 2 & y \end{bmatrix} \otimes \begin{bmatrix} 1 & y \\ 2 & x \end{bmatrix}.$$

25. Find the trace of

$$\begin{bmatrix} 3 & 8 \\ 7 & 9 \end{bmatrix} \otimes \begin{bmatrix} 8 & 6 \\ 5 & 9 \end{bmatrix}$$

over $(\mathbb{I}_{23}, +, \cdot)$.

CHAPTER III
Fourier Evaluation and Interpolation

1 Introduction

The topic of discussion on error-free computation will be incomplete if at
this point we do not introduce the closely related concepts of error-free
Fourier transforms and convolutions based on modular computations.

In this chapter we will study the relationship between:

(i)   Fourier transform and polynomial evaluation;
(ii) inverse Fourier transform and polynomial interpolation; and
(iii) convolution and polynomial multiplication.

We also will describe how exact computational methods are used for this
purpose.

2 Discrete Fourier Transform over a Ring

The Fourier transform is usually defined over the complex numbers.
However, here, we will define the Fourier transform over an arbitrary
commutative ring $(R, +, \cdot, 0, 1)$.

2.1 Primitive nth root of unity

An element ω of R such that

$$\omega \neq 1$$
$$\omega^n = 1$$

and

$$\sum_{j=0}^{n-1} \omega^{jp} = 0, \quad \text{for } 1 \le p \le n-1$$

is said to be the principal or primitive nth root of unity. The elements $\omega^0, \omega^1, \ldots, \omega^{n-1}$ are the nth roots of unity.

2.1.1 EXAMPLE. $e^{2\pi i/n}$, where $i = \sqrt{-1}$ is a principal nth root of unity in the ring of complex numbers.

2.2 Discrete Fourier transform (DFT)

Let $P^T = [p_0, p_1, \ldots, p_{n-1}]$ be an $(1 \times n)$ vector (or a vector of length n) with elements from R. We assume that the integer n has a multiplicative inverse in R and that R has a principal nth root of unity ω.

Let A be an $(n \times n)$ matrix with $a_{ij} = \omega^{ij}, 0 \le i, j \le (n-1)$. The vector $AP = F(P) = b$ whose ith component $b_i, 0 \le i \le (n-1)$ is

$$\sum_{k=0}^{n-1} p_k \omega^{ik}$$

is called the discrete Fourier transform (DFT) of P.

We notice that $\sum_{k=0}^{n-1} p_k \omega^{ik}$ corresponds to the evaluation of the polynomial

$$\mathscr{P}(x) = \sum_{k=0}^{n-1} p_k x^k$$

at the point $x = \omega^i$.

Thus the components b_i $(0 \le i \le n-1)$ correspond to the evaluation of the polynomial $\mathscr{P}(x)$ at the points $x = \omega^0, x = \omega^1, \ldots, x = \omega^{n-1}$. It is also to be noted that the matrix A is a particular type of Vandermonde matrix (Young and Gregory [1973]), called "Fourier matrix":

$$A = \begin{bmatrix} (\omega^0)^0 & (\omega^0)^1 & \cdots & (\omega^0)^{n-1} \\ (\omega^1)^0 & (\omega^1)^1 & \cdots & (\omega^1)^{n-1} \\ \cdot & \cdot & \cdot & \cdot \\ \cdot & \cdot & \cdot & \cdot \\ \cdot & \cdot & \cdot & \cdot \\ (\omega^{n-1})^0 & \cdot & \cdots & (\omega^{n-1})^{n-1} \end{bmatrix}$$

2.3 **Theorem.** *The matrix A has the following two properties. (Aho, Hopcroft, and Ullman [1975], Lipson [1981]).*

Property 1: A is nonsingular;
Property 2: A^{-1} has as its ijth element $(1/n)\omega^{-ij} = n^{-1}\omega^{-ij}$, where $n^{-1} \in R$.

PROOF. Property 1 is easily proved, since A is a Vandermonde matrix with distinct elements.

To prove Property 2, let us replace a_{ij} by a_{ij}^{-1} and call this new matrix B. If we show that

$$AB = nI = \begin{bmatrix} n & 0 & \cdots & 0 \\ 0 & n & \cdots & 0 \\ 0 & \cdot & \cdots & n \end{bmatrix},$$

where I is the identity matrix of order $(n \times n)$ then our proof is complete.

Let $AB = C$; then

$$c_{ij} = \sum_{k=0}^{n-1} \omega^{ik} \omega^{-kj} \qquad (0 \leq i, j < n);$$

for $i = j, \qquad c_{ii} = n$

for $i \neq j, \qquad c_{ij} = \sum_{k=0}^{n-1} \omega^{(i-j)k}.$

Since $|(i-j)| < n$, $\omega^{i-j} \neq 1$. Therefore, $\sum_{k=0}^{n-1} \omega^{(i-j)}$ can be summed up using the geometric series formula to give

$$\sum_{k=0}^{n-1} \omega^{(i-j)k} = \frac{(\omega^{i-j})^n - 1}{(\omega^{i-j} - 1)} = 0. \qquad \square$$

2.4 Inverse discrete Fourier transform (IDFT)

We define

$$A^{-1}F(P) = P$$

or

$$p_i = \frac{1}{n} \sum_{k=0}^{n-1} \mathcal{P}(\omega^k) \omega^{-ik}$$

as the inverse discrete Fourier transform (IDFT) of $[\mathcal{P}(\omega^0), \ldots, \mathcal{P}(\omega^{n-1})]$. Note that this operation performs the Fourier interpolation of finding the coefficients of $\mathcal{P}(x)$ given the values of $\mathcal{P}(x_i)$ at the n points x_i.

2.4.1 EXAMPLE. Let $P^T = [1, 2, 3, 4]$ over $(\mathbb{I}_{17}, +, \cdot)$; we illustrate how to perform DFT of P and IDFT of P.

Since $n = 4$ we choose $\omega = 4$ which is the principal 4th root of unity, viz., $\omega^4 = 256 \equiv 1 \bmod 17$, we have now $\omega^0 = 1$, $\omega^1 = 4$, $\omega^2 = 16$, $\omega^3 = 13$ as the 4th roots of unity.

$$\text{The matrix } A = \begin{bmatrix} 1 & 1 & 1 & 1 \\ 1 & 4 & 4^2 & 4^3 \\ 1 & 4^2 & 4^4 & 4^6 \\ 1 & 4^3 & 4^6 & 4^9 \end{bmatrix} \bmod 17$$

$$= \begin{bmatrix} 1 & 1 & 1 & 1 \\ 1 & 4 & 16 & 13 \\ 1 & 16 & 1 & 16 \\ 1 & 13 & 16 & 4 \end{bmatrix}.$$

The DFT of P

$$AP = \begin{bmatrix} 1 & 1 & 1 & 1 \\ 1 & 4 & 16 & 13 \\ 1 & 16 & 1 & 16 \\ 1 & 13 & 16 & 4 \end{bmatrix} \begin{bmatrix} 1 \\ 2 \\ 3 \\ 4 \end{bmatrix} = \begin{bmatrix} 10 \\ 7 \\ 15 \\ 6 \end{bmatrix}.$$

We have

$$A^{-1} = \frac{1}{4} \begin{bmatrix} 1 & 1 & 1 & 1 \\ 1 & 4^{-1} & 4^{-2} & 4^{-3} \\ 1 & 4^{-2} & 4^{-4} & 4^{-6} \\ 1 & 4^{-3} & 4^{-6} & 4^{-9} \end{bmatrix} = 13 \begin{bmatrix} 1 & 1 & 1 & 1 \\ 1 & 13 & 16 & 4 \\ 1 & 16 & 1 & 16 \\ 1 & 4 & 16 & 13 \end{bmatrix}.$$

It is easily seen $F^{-1}F(P) = A^{-1}AP = P$.

3 Convolution

We now define the convolution of two sequences each of length n. Let $p = [p_0, p_1, \ldots, p_{n-1}]$ and $q = [q_0, q_1, \ldots, q_{n-1}]$ be two sequences each of length n. The convolution of p and q is denoted by $p * q$.

$p * q$ is the $2n$ length sequence $r = [r_0, r_1, \ldots, r_{2n-1}]$ where

(3.1)
$$r_i = \sum_{j=0}^{n-1} p_j q_{i-j}$$

(where p_k and q_k are assumed to be zero for $k < 0$ and $k \geq n$; otherwise, we obtain the cyclic convolution, where the sequences are assumed to repeat after n).

Thus

$$r_0 = p_0 q_0$$
$$r_1 = p_0 q_1 + p_1 q_0$$
$$r_2 = p_0 q_2 + p_1 q_1 + p_2 q_0$$

(3.2)
$$\vdots$$

$$r_{2n-3} = p_{n-2} q_{n-1} + p_{n-1} q_{n-2}$$
$$r_{2n-2} = p_{n-1} q_{n-1}$$
$$r_{2n-1} = 0.$$

3.1 Cyclic convolution property (CCP)

A DFT is said to possess the cyclic convolution property if

$$(3.3) \qquad\qquad DFT(p * q) = DFT(p) \cdot DFT(q)$$

where $\cdot$ denotes pointwise (componentwise) product. From (3.1) and (3.3) we can compute a convolution thus:

$$(3.4) \qquad\qquad r = p * q = IDFT\{DFT(p) \cdot DFT(q)\}.$$

Equation (3.4) for computing convolution can be realized by the following three-step procedure which involves two DFT, one IDFT, and n multiplications:

Step 1: Find the DFT of p and q;
Step 2: Multiply $DFT(p)$ and $DFT(q)$ componentwise;
Step 3: Inverse transform (IDFT) the product obtained in Step 2.

Equation (3.4) is called cyclic convolution, since it evaluates (3.4), as if p and q were periodically extended outside the range 0 to $(n-1)$ or equivalently, the indices are evaluated modulo n. Normal finite convolution can be calculated by cyclic convolution, if zeros are appended to p and q to prevent folding. In other words, since the original sequences are of length n and the convolution is of length $2n$, the two original sequences are extended to a length $2n$ by appending n zeros. This introduction of additional or guarding zeros is called "padding."

3.2 Remarks. The use of Fourier transforms to compute convolution became useful only when Cooley and Tukey [1961] introduced the famous (now classical) algorithm known as the Fast Fourier transform (FFT) to compute the Fourier transform in $O(n \log n)$ steps rather than $O(n^2)$ steps. For details, see Rabiner and Gold [1976] and Nussbaumer [1980], Gold and Rader [1969], Rabiner and Rader [1972], Henrici [1979], Borodin and Munro [1975], Merz [1983].

4 Error-Free DFT

It is well known that DFT is one of the most fundamental operations in digital signal processing and many other areas. Its importance is seen from the enormous amount of research effort that has gone into this area.

The computation of Fourier transform of a sequence of length n requires n^2 complex multiplications and additions. However, using the classical and most powerful Cooley–Tukey FFT algorithm the number of complex multiplications and additions can be brought down to $(n/2) \log_2 n$ when n is a power of two.

For example, if $n = 2^{10} = 1024$ the FFT algorithm is faster by a factor of 200. Efficient application of FFT algorithm requires that n be highly composite, preferably a power of 2 and in most practical problems, this restriction can be easily imposed.

When we work with finite precision complex numbers, the round-off errors accumulate to a large extent. One of the essential aspects in digital signal processing is to eliminate the round-off errors in DFT. Attempts towards this have been made by several authors; Agarwal and Burrus [1975], Pollard [1971], Rader [1972], McClellan and Rader [1980], Nussbaumer [1980], Reed and Truong [1975]. These error-free methods are also called "Number theoretic transform methods" and "Polynomial transform methods." Here we will deal only with the "Number-theoretic transform" methods; for "Polynomial transform," see Nussbaumer [1980], Winograd [1978]. Our aim here is only to provide a relevant summary of the results in this area, since several books are already available (McClellan and Rader [1980], Nussbaumer [1980]).

4.1 Basic number theoretic results

In order to describe the number theoretic transform methods we need to introduce the following standard definitions and theorems from number theory (Apostol [1976], Le Veque [1977], MacDuffee [1948]).

4.1.1 *Exponent of an integer* mod m. Let a and m be relatively prime integers with $m \geq 1$. Consider all the positive powers of a:

$$a, a^2, \ldots, a^f.$$

The smallest positive integer f such that

$$a^f \equiv 1 \bmod m$$

is called the exponent of a mod m or order of a with respect to m and is denoted by

$$f = \exp_m(a).$$

4.1.2 *Euler's totient function* $\phi(m)$. If $m \geq 1$, the Euler's totient function is defined to be the number of positive integers not exceeding m which are relatively prime to m; it is well known that (Apostol [1976], Hardy and Wright [1979])

$$\phi(m) = m \prod_{i=1}^{k} (1 - 1/p_i),$$

where $m = \prod_{i=1}^{k} p_i^{r_i}$, the prime decomposition of m. If p is prime and $r \geq 1$, then $\phi(p^r) = p^r - p^{r-1}$.

4.1.3 *Primitive root.* If $\exp_m \omega = \phi(m)$, then ω is called a primitive root modulo m.

4.1.4 Theorem (Euler). *If* $\gcd(a,m) = 1$, *then* $a^{\phi(m)} \equiv 1 \bmod m$.

4.1.5 Theorem (Fermat). *If* p *is a prime* $a^{p-1} \equiv 1 \bmod p$.

4.1.6 Theorem. *If* $\omega^n \equiv 1 \bmod m$, *then* n *must divide* $\phi(m)$.

4.1.7 Theorem. *If* m *is a prime, then roots of order* n *exist if and only if* n *divides* $(m-1)$ *and these roots are given by*

$$\omega = \omega_\phi^{(m-1)/n},$$

where ω_ϕ *is a primitive root.*

4.1.8 Theorem. *If* ω *is a root of order* n *then* ω^k *is of order* n/k *if* k *divides* n; ω^k *is of order* n, *if* $\gcd(n,k) = 1$.

4.2 Number theoretic transforms (NTT)

Discrete Fourier transforms in which the complex exponentials are replaced by the primitive roots and with operations performed modulo an integer M (prime or composite) are known as number theoretic transforms. The following important theorems classify their properties (see Nussbaumer [1980]).

4.2.1 Theorem. *A NTT of length* n *and primitive root* ω *possesses cyclic convolution property when defined modulo a prime* p, *if and only if* ω *is a root of unity of order* n *modulo* p; *i.e., the least* n *such that* $\omega^n \equiv 1 \bmod p$.

4.2.2 Theorem. *A NTT of length* n *defined modulo* p *(a prime) exists if and only if* n *divides* $(p-1)$. *This NTT has cyclic convolution property* (CCP).

Theorems 4.2.1 and 4.2.2 specify the size of the NTT for a given p. NTT is defined completely, provided we can find a root of order n.

4.2.3 Theorem. *A NTT of length* n *and root* ω *defined modulo a composite integer* M, *possesses CCP if and only if*

$$\omega^n \equiv 1 \bmod M;$$

$$n \cdot n^{-1} \equiv 1 \bmod M;$$

and $\gcd(\omega^d - 1, M) = 1$, *for every* d *such that* n/d *is a prime.*

This is Erdelky's theorem (see Nussbaumer [1980]).

4.2.4 Remark. For $M = p_1^{r_1}$, Theorem 4.2.3 implies that n divides $(p_1 - 1)$;

this is so since n^{-1} exists only if n and M are relatively prime; by Theorem 4.1.6, n must divide $\phi(M) = p_1^{r_1-1}(p_1 - 1)$ (by 4.1.2). Therefore n divides $(p_1 - 1)$.

4.2.5 Theorem. *For $M = \prod_{i=0}^{k} p_i^{r_i}$, a length n NTT defined mod M supports cyclic convolution if and only if*

$$n \mid \gcd\{(p_1 - 1), (p_2 - 1), \ldots, (p_k - 1)\}.$$

The maximum transform length is $\gcd\{(p_1 - 1), \ldots, (p_k - 1)\}$.

4.2.6 Remark. For NTT's to be attractive in comparison to other implementations of convolution, they should be computationally efficient. In order to have a fast FFT type algorithm, n should be highly composite, preferably a power of 2 and should be large enough for practical sequence lengths. Also, the multiplication operation by powers of ω should be a simple operation. If ω is 2 or a power of 2, the multiplication reduces to simple word shifts. In addition, the arithmetic modulo M should be simple, but large enough to handle the practical problems. Thus the main problem in using NTT for evaluation–interpolation is to identify M and the resulting n and ω. With this in mind several transforms have been thought of. We will mention here, the Mersenne, Fermat, and related Rader transforms.

4.3 Mersenne transforms

If M is even, it has a factor 2 and so $n_{\max} = 1$. Thus to be of practical use, M has to be odd. If M is a prime, then $n_{\max} = M - 1$ which is the maximum possible length. When $M = 2^p - 1$, where p is an odd prime, M is called a Mersenne number. Number-theoretic transforms with Mersenne number as modulus are called Mersenne transforms (Rader [1972]); for such values of M, the integer $\omega = 2$ is a root of order p thus giving a length $n = p$. Also $\omega = -2$ gives a transform of length $n = 2p$. The advantage is that these transforms can be realized by shift operations instead of multiplications. However, the principal deficiencies are the lack of a fast transform and the rigid relationship between word length M and transform length n.

4.4 Fermat and Rader transforms

Let $M = F_t = 2^{2^t} + 1 = 2^b + 1$, where $b = 2^t$, t any positive integer; M is called a Fermat number. The first few Fermat numbers are:

$$F_0 = 3, \qquad F_1 = 5, \qquad F_2 = 17, \qquad F_3 = 257, \qquad F_4 = 65537$$

which are also primes. However, beyond F_5 the Fermat numbers could be composite.

Number theoretic transforms which use Fermat numbers as moduli are called Fermat transforms (Agarwal and Burrus [1975]).

Since Fermat numbers are primes up to F_4, the maximum possible transform length for $M = 2^b + 1$ is $n_{max} = 2^b$ and we can have a Fermat transform for any length $n = 2^m$, $m \leqslant b$. For these Fermat primes, 3 is an ω of order 2^b allowing the largest possible transform length for $t \geqslant 1$. The integer 2 is of order $n = 2b = 2^{t+1}$. If ω is taken as 2, or a power of 2, all the powers of ω would be some powers of 2; for these cases the Fermat transforms can be computed very efficiently. These transforms are called Rader transforms. Since n is a power of 2, the FFT-like algorithm can be used and multiplications are realized as shift operations.

5 Polynomial Evaluation–Interpolation– Multiplication

As already mentioned, there is a close relationship between Fourier transforms and polynomial evaluations.

Consider, for instance, an $(n-1)$ degree polynomial

$$(5.1) \qquad \mathscr{P}(x) = \sum_{i=0}^{n-1} p_i x^i.$$

This polynomial can be represented either by a list of its coefficients $[p_0, p_1, \ldots, p_{n-1}]$ or by a list of its values evaluated at n distinct points $[x_0, x_1, \ldots, x_{n-1}]$.

Interpolation is the process of finding the coefficient representation of the polynomials given the values at the n distinct points $[x_0, x_1, \ldots, x_{n-1}]$. From the definition of DFT, it is clear that the sequence $[\mathscr{P}(\omega^0), \mathscr{P}(\omega^1), \ldots, \mathscr{P}(\omega^{n-1})]$ is just the values of $\mathscr{P}(x)$ evaluated at $[\omega^0, \omega^1, \ldots, \omega^{n-1}]$ or the DFT performs the evaluation of $\mathscr{P}(x)$.

The IDFT on the other hand defined by

$$(5.2) \qquad p_i = \frac{1}{n} \sum_{k=0}^{n-1} \mathscr{P}(\omega^k) \omega^{-ik}$$

actually interpolates the polynomial or finds the coefficients p_i given the values of the function at $\omega^0, \omega^1, \ldots, \omega^{n-1}$.

For our purpose (and more so for generalization) it is convenient to represent (see Chapter II, Section 3)

$$(5.3) \qquad \mathcal{P}(x) = [1 \, x \, x^2 \cdots x^{n-1}] \begin{bmatrix} p_0 \\ p_1 \\ \vdots \\ p_{n-1} \end{bmatrix}$$

$$= X^T P.$$

where

$$P^T = [\, p_0 p_1 \cdots p_{n-1}]$$
$$X^T = [1 \, x \, x^2 \cdots x^{n-1}].$$

Thus $\mathcal{P}(x_i)$ gives the evaluation of $\mathcal{P}(x)$ at x_i if the values of x_i are placed in (5.3). On the other hand, if $\mathcal{P}(x_i)$ are known then the vector P is computed using

$$(5.4) \qquad P = (X_i^T)^{-1} \mathcal{P}(x_i)$$

(readers are referred to the Chapter II, Section 3, for the notation), where

$$X_i^T = \begin{bmatrix} 1 & x_0 & x_0^2 & \cdots & x_0^{n-1} \\ \vdots & \vdots & \vdots & & \vdots \\ 1 & x_{n-1} & x_{n-1}^2 & \cdots & x_{n-1}^{n-1} \end{bmatrix}$$

and

$$[\mathcal{P}(x_i)]^T = [\mathcal{P}(x_0) \, \mathcal{P}(x_1) \cdots \mathcal{P}(x_{n-1})].$$

In Chapter II we explained that the polynomial matrix inversion requires evaluation and interpolation of polynomials and polynomial multiplications over a field. It is clear that these can be achieved using the DFT and convolutions. With this as our aim, we now explain methods based on number theoretic transform for polynomial multiplication.

5.1 Single-variable polynomial multiplication

Consider the product of two $(n-1)$ degree polynomials

$$\mathcal{P}(x) = \sum_{i=0}^{n-1} p_i x^i$$

$$\mathcal{Q}(x) = \sum_{i=0}^{n-1} q_i x^i.$$

The product

$$\mathcal{P}(x) \cdot \mathcal{Q}(x) = \mathcal{R}(x) = \sum_{i=0}^{2n-2} \sum_{j=0}^{i} p_j q_{i-j} x^i$$

has the coefficients which are exactly the components of the convolution of the sequences

$$P^T = [p_0, p_1, \ldots, p_{n-1}] \quad \text{and} \quad Q^T = [q_0, q_1, \ldots, q_{n-1}]$$

and hence can be realized by the use of number theoretic transforms with a high speed. However, as mentioned earlier (Section 3) we need to pad each of these sequences P^T and Q^T with n additional zeros so that the resulting sequence of length $2n$ fully represents the product.

5.2 EXAMPLE. Multiply the two polynomials

$$\mathcal{P}(x) = 3 + 2x$$

$$\mathcal{Q}(x) = 6 + 7x$$

and obtain their product $\mathcal{R}(x)$ over $(\mathbb{I}_{17}, +, \cdot)$ using DFT, IDFT through NTT.

Since $M = 17$ and $n \mid (M-1)$, we choose $\omega = 4$. We are given the two sequences (appended with zeros)

$$P^T = [3, 2, 0, 0]; \qquad Q^T = [6, 7, 0, 0].$$

Thus the evaluated values of $\mathcal{P}(x_i)$ and $\mathcal{Q}(x_i)$ are

$$\mathcal{P}(x_i) = X_i^T P = \begin{bmatrix} 1 & 1 & 1 & 1 \\ 1 & 4 & 16 & 13 \\ 1 & 16 & 1 & 16 \\ 1 & 13 & 16 & 4 \end{bmatrix} \begin{bmatrix} 3 \\ 2 \\ 0 \\ 0 \end{bmatrix} = \begin{bmatrix} 5 \\ 11 \\ 1 \\ 12 \end{bmatrix} = \begin{bmatrix} \mathcal{P}(1) \\ \mathcal{P}(4) \\ \mathcal{P}(16) \\ \mathcal{P}(13) \end{bmatrix}$$

$$\mathcal{Q}(x_i) = X_i^T Q = \begin{bmatrix} 1 & 1 & 1 & 1 \\ 1 & 4 & 16 & 13 \\ 1 & 16 & 1 & 16 \\ 1 & 13 & 16 & 4 \end{bmatrix} \begin{bmatrix} 6 \\ 7 \\ 0 \\ 0 \end{bmatrix} = \begin{bmatrix} 13 \\ 0 \\ 16 \\ 12 \end{bmatrix} = \begin{bmatrix} \mathcal{Q}(1) \\ \mathcal{Q}(4) \\ \mathcal{Q}(16) \\ \mathcal{Q}(13) \end{bmatrix}$$

$$\mathcal{R}(x_i) = (X_i^T P \cdot X_i^T Q) = \begin{bmatrix} 14 \\ 0 \\ 16 \\ 8 \end{bmatrix} = \begin{bmatrix} \mathcal{P}(1) \cdot \mathcal{Q}(1) \\ \mathcal{P}(4) \cdot \mathcal{Q}(4) \\ \mathcal{P}(16) \cdot \mathcal{Q}(16) \\ \mathcal{P}(13) \cdot \mathcal{Q}(13) \end{bmatrix} = \begin{bmatrix} \mathcal{R}(1) \\ \mathcal{R}(4) \\ \mathcal{R}(16) \\ \mathcal{R}(13) \end{bmatrix}.$$

Thus

$$R = (X_i^T)^{-1} \mathcal{R}(x_i) = 13 \begin{bmatrix} 1 & 1 & 1 & 1 \\ 1 & 13 & 16 & 4 \\ 1 & 16 & 1 & 16 \\ 1 & 4 & 16 & 13 \end{bmatrix} \begin{bmatrix} 14 \\ 0 \\ 16 \\ 8 \end{bmatrix} = \begin{bmatrix} 1 \\ 16 \\ 14 \\ 0 \end{bmatrix} = \begin{bmatrix} r_0 \\ r_1 \\ r_2 \\ r_3 \end{bmatrix}$$

and $\mathcal{R}(x) = 1 + 16x + 14x^2 = \mathcal{P}(x) \cdot \mathcal{Q}(x)$.

5.3 Choice of modulus for computation

If the coefficients of $\mathscr{P}(x)$ and $\mathscr{Q}(x)$ are integers then in order to apply modular arithmetic for convolution and DFT, the coefficient r_j in $\mathscr{R}(x)$ must satisfy the condition

$$-M/2 < r_j < M/2$$

in order to avoid overflow of the coefficients; here M is the modulus used for computation. If $|p_i| \leqslant \alpha$ and $|q_i| \leqslant \beta$ then $|r_j|_{\max} = n \, |\alpha\beta| < M/2$ or

(5.5) $$M > 2n \, |\alpha\beta|$$

so that the result r_j does not overflow.

If the coefficients p_i and q_i are very large, computations can be carried out using many suitable smaller primes and the results can be combined using the Chinese remainder theorem (Chapter I); Theorem 4.2.5 enables us to choose n.

5.4 EXAMPLE. We now give another example which illustrates the use of Mersenne transform. Let $|\alpha|, |\beta| \leqslant 2$, thus $\mathscr{P}(x) = 2 + 2x + x^2$, $\mathscr{Q}(x) = 2 + x + x^2$;

$$P^T = [2, 2, 1, 0, 0]; \qquad Q^T = [2, 1, 1, 0, 0].$$

We choose $M = 2^5 - 1 = 31$ and $\omega = 2$, giving a length $n = 5$. Thus the $\mathscr{P}(x_i)$ and $\mathscr{Q}(x_i)$ are given by:

$$\mathscr{P}(x_i) = X_i^T P = \begin{bmatrix} 1 & 1 & 1 & 1 & 1 \\ 1 & 2 & 4 & 8 & 16 \\ 1 & 4 & 16 & 2 & 8 \\ 1 & 8 & 2 & 16 & 4 \\ 1 & 16 & 8 & 4 & 2 \end{bmatrix} \begin{bmatrix} 2 \\ 2 \\ 1 \\ 0 \\ 0 \end{bmatrix} = \begin{bmatrix} 5 \\ 10 \\ 26 \\ 20 \\ 11 \end{bmatrix},$$

$$\mathscr{Q}(x_i) = X_i^T Q = \begin{bmatrix} 1 & 1 & 1 & 1 & 1 \\ 1 & 2 & 4 & 8 & 16 \\ 1 & 4 & 16 & 2 & 8 \\ 1 & 8 & 2 & 16 & 4 \\ 1 & 16 & 8 & 4 & 2 \end{bmatrix} \begin{bmatrix} 2 \\ 1 \\ 1 \\ 0 \\ 0 \end{bmatrix} = \begin{bmatrix} 4 \\ 8 \\ 22 \\ 12 \\ 26 \end{bmatrix},$$

$$\mathscr{P}(x_i) \cdot \mathscr{Q}(x_i) = \begin{bmatrix} 20 \\ 18 \\ 14 \\ 23 \\ 7 \end{bmatrix} = \begin{bmatrix} \mathscr{R}(1) \\ \mathscr{R}(2) \\ \mathscr{R}(4) \\ \mathscr{R}(8) \\ \mathscr{R}(16) \end{bmatrix} = \mathscr{R}(x_i),$$

$$R = (X_i^T)^{-1} \mathcal{R}(x_i) = \frac{1}{5} \begin{bmatrix} 1 & 1 & 1 & 1 & 1 \\ 1 & 16 & 8 & 4 & 2 \\ 1 & 8 & 2 & 16 & 4 \\ 1 & 4 & 16 & 2 & 8 \\ 1 & 2 & 4 & 8 & 16 \end{bmatrix} \begin{bmatrix} 20 \\ 18 \\ 14 \\ 23 \\ 7 \end{bmatrix} = \frac{1}{5} \begin{bmatrix} 20 \\ 30 \\ 30 \\ 15 \\ 5 \end{bmatrix} = \begin{bmatrix} 4 \\ 6 \\ 6 \\ 3 \\ 1 \end{bmatrix}$$

Thus $\mathcal{R}(x) = 4 + 6x + 6x^2 + 3x^3 + x^4$.

5.5 Complexity reduction

We will now briefly indicate the gain in speed that would result in using FFT based NTT and convolution for polynomial multiplication.

If $\mathcal{P}(x)$ and $\mathcal{Q}(x)$ are both polynomials of degree $(n-1)$ then the direct product requires $\mu_D = n^2$ multiplications. However, if we use FFT based DFT and number theoretic transform, it will take $\mu_C = 3 \cdot (2n/2) \cdot \log_2 2n + 2n$ multiplications, since convolution requires 3 DFT and a direct product of $2n$ component vectors. Thus

$$\frac{\mu_D}{\mu_C} = \frac{n}{3 \log_2 n + 5}.$$

Table 5.1 shows that only for large n this method would be economical.

Table 5.1   Complexity reduction

n	μ_D/μ_C
16	0.9
32	1.6
64	2.8
128	5.0

6 Multivariable Polynomial Interpolation

The procedure above can be easily extended to the case of multivariable polynomials, using the notion of tensor product described in Chapter II, Section 4. For the case of multivariable polynomials with a large number of variables each of a high degree, the FFT method is very effective.

A polynomial $\mathcal{P}(x_1, x_2, \ldots, x_m)$ in m variables can be written as

(6.1) $$\mathcal{P}(x_1, x_2, \ldots, x_m) = \sum_{i_1=0}^{n-1} \sum_{i_2=0}^{n-1} \cdots \sum_{i_m=0}^{n-1} p_{i_1 i_2 \cdots i_m} x_1^{i_1} \cdots x_m^{i_m}.$$

Consider the product of two m-variable polynomials $\mathcal{P}(x_1, \ldots, x_m)$ and $\mathcal{Q}(x_1, \ldots, x_m)$ defined as above (6.1). Direct multiplication of $\mathcal{P}$ and $\mathcal{Q}$ would require n^{2m} multiplications. However, using an m-dimensional FFT algorithm the number of multiplications can be drastically reduced.

An m-dimensional algorithm is applied to the evaluation and interpolation in the following manner using the tensor product scheme described in the last chapter. We evaluate $\mathcal{P}(x_1, x_2, \ldots, x_m)$ using the analogue of (4.8) of Chapter II, viz.

(6.2) $\qquad v(\mathcal{P}(x_1, x_2, \ldots, x_m)) = (X_m^T \otimes \cdots \otimes X_1^T) v(P^T).$

As in the case of one variable example (Section 5.1) we need to append or pad n additional zeros to each one of the vectors to take care of the increase in the degree of each variable after a multiplication. Thus we can estimate the length of $v(R^T)$ as $(2n)^m$.

Thus to compute the product of two polynomials $\mathcal{P}$ and $\mathcal{Q}$ we evaluate $v(\mathcal{P})$ and $v(\mathcal{Q})$ using (6.2) and multiply these vectors componentwise to obtain $v(\mathcal{R})$. Then the interpolation of $\mathcal{R}$ is carried out using the analogue of (4.7) of Chapter II, viz.,

(6.3) $\qquad v(R^T) = [(X_m^T)^{-1} \otimes \cdots \otimes (X_1^T)^{-1}] v(\mathcal{R}(x_1, \ldots, x_m)),$

where $v(\mathcal{R}) = v(\mathcal{P}) \cdot v(\mathcal{Q})$.

6.1 EXAMPLE. Let

$$\mathcal{P}(x,y) = 1 + 2x$$
$$\mathcal{Q}(x,y) = 2 + y.$$

Find $\mathcal{R}(x,y)$ using DFT and NTT over $(\mathbb{I}_{17}, +, \cdot)$.

In this case, for simplicity we choose $\omega = 16$ since we know that $v(R)$ has a length 4. (However, note that for the general case of two polynomials in two variables each of degree 1, we need to choose ω such that $n = 16$.)

Thus

$$X_i^T = \begin{bmatrix} 1 & 1 \\ 1 & 16 \end{bmatrix}$$

$$Y_j^T = \begin{bmatrix} 1 & 1 \\ 1 & 16 \end{bmatrix}$$

and

$$v(\mathcal{P}(x_i,y_j)) = \left[\begin{bmatrix} 1 & 1 \\ 1 & 16 \end{bmatrix} \otimes \begin{bmatrix} 1 & 1 \\ 1 & 16 \end{bmatrix} \right] \begin{bmatrix} 1 \\ 2 \\ 0 \\ 0 \end{bmatrix} = \begin{bmatrix} 3 \\ 16 \\ 3 \\ 16 \end{bmatrix}$$

$$v(\mathcal{Q}(x_i,y_j)) = \left[\begin{bmatrix} 1 & 1 \\ 1 & 16 \end{bmatrix} \otimes \begin{bmatrix} 1 & 1 \\ 1 & 16 \end{bmatrix} \right] \begin{bmatrix} 2 \\ 0 \\ 1 \\ 0 \end{bmatrix} = \begin{bmatrix} 3 \\ 3 \\ 1 \\ 1 \end{bmatrix}$$

$$v(\mathcal{P}) \cdot v(\mathcal{Q}) = \begin{bmatrix} 9 \\ 14 \\ 3 \\ 16 \end{bmatrix} = v(\mathcal{R}(x_i, y_j))$$

$$v(R^T) = \left[\begin{bmatrix} 1 & 1 \\ 1 & 16 \end{bmatrix}^{-1} \otimes \begin{bmatrix} 1 & 1 \\ 1 & 16 \end{bmatrix}^{-1} \right] \begin{bmatrix} 9 \\ 14 \\ 3 \\ 16 \end{bmatrix}$$

$$= \left[\frac{1}{2} \begin{bmatrix} 1 & 1 \\ 1 & 16 \end{bmatrix} \otimes \frac{1}{2} \begin{bmatrix} 1 & 1 \\ 1 & 16 \end{bmatrix} \right] \begin{bmatrix} 9 \\ 14 \\ 3 \\ 16 \end{bmatrix}$$

$$= 13 \begin{bmatrix} 1 & 1 & 1 & 1 \\ 1 & 16 & 1 & 16 \\ 1 & 1 & 16 & 16 \\ 1 & 16 & 16 & 1 \end{bmatrix} \begin{bmatrix} 9 \\ 14 \\ 3 \\ 16 \end{bmatrix} = \begin{bmatrix} 2 \\ 4 \\ 1 \\ 2 \end{bmatrix}$$

or $\mathcal{R}(x,y) = 2 + 4x + y + 2xy$.

6.2 Choice of resulting vector length and coefficient size

We already mentioned that in the m variable case where each vector is of length n, we choose the length of the result to be $(2n)^m$, and pad n additional zeros to each vector.

If the coefficients of $\mathcal{P}$ and $\mathcal{Q}$ are (real) integers then in order to apply convolution and DFT using the modular arithmetic, M has to be chosen so that the product $\mathcal{R}$ has coefficients in the range

$$-M/2 < r_{ij} < M/2,$$

where M is the modulus used for computation. If $|\alpha|$ and $|\beta|$ are the maximal coefficients in $\mathcal{P}$ and $\mathcal{Q}$ then the product $\mathcal{R}$ can have a maximum coefficient size

(6.4) $|r_{ij}| < n^m |\alpha\beta|;$

thus we need to choose $M > 2n^m |\alpha\beta|$. If M is large, one can choose several primes p_i and combine the results using the Chinese remainder theorem and related procedures; we must, however, make a proper choice of the vector length using Theorem 4.2.5.

6.3 EXAMPLE. Let

$$\mathcal{P}(x,y) = \mathcal{Q}(x,y)$$
$$= (1 + 2x + 3x^2) + (2 + 3x + 4x^2)y + (3 + 4x + 5x^2)y^2.$$

Find the product $\mathcal{P} \cdot \mathcal{Q}$ by Fourier evaluation and interpolation. For this example we simply use the estimate $M = 257$ and length 64 as these are adequate (note that the estimate of M given in Section 6.2 is rather large).

We thus have $\omega = 4$ and $4^8 \equiv 1 \mod 257$. The Fourier matrices X_i^T and Y_j^T are (8×8) matrices with $\omega = 4$.

We now proceed as in Example 6.1 to evaluate $v(\mathscr{P}(x_i, y_j))$ and $v(\mathscr{Q}(x_i, y_j))$ and convolve these to obtain $v(\mathscr{R}(x_i, y_j))$. Note that all these vectors are of length 64, but conveniently expressed as (8×8) matrices $\mathscr{P}(x_i, y_j)$, $\mathscr{Q}(x_i, y_j)$ below

$$\mathscr{P}(x_i, y_j) = \mathscr{Q}(x_i, y_j) = \begin{bmatrix} 27 & 234 & 138 & 133 & 9 & 162 & 107 & 9 \\ 234 & 154 & 178 & 124 & 78 & 44 & 252 & 163 \\ 138 & 178 & 190 & 183 & 46 & 67 & 3 & 206 \\ 133 & 124 & 183 & 140 & 130 & 112 & 135 & 205 \\ 9 & 78 & 46 & 130 & 3 & 54 & 207 & 3 \\ 162 & 44 & 67 & 112 & 54 & 126 & 138 & 139 \\ 107 & 252 & 3 & 135 & 207 & 138 & 61 & 110 \\ 9 & 163 & 206 & 205 & 3 & 139 & 110 & 78 \end{bmatrix}$$

Computing the direct product $v(\mathscr{R}) = v(\mathscr{P}) \cdot v(\mathscr{Q})$ and using

$$v(R^T) = [(Y_j^T)^{-1} \otimes (X_i^T)^{-1}] v(\mathscr{R}(x_i, y_j))$$

we obtain $v(R^T)$ as a (64×1) vector whose R^T matrix is given by

$$R^T = \begin{bmatrix} 1 & 4 & 10 & 12 & 9 & 0 & 0 & 0 \\ 4 & 14 & 32 & 34 & 24 & 0 & 0 & 0 \\ 10 & 32 & 69 & 68 & 46 & 0 & 0 & 0 \\ 12 & 34 & 68 & 62 & 40 & 0 & 0 & 0 \\ 9 & 24 & 46 & 40 & 25 & 0 & 0 & 0 \\ 0 & 0 & 0 & 0 & 0 & 0 & 0 & 0 \\ 0 & 0 & 0 & 0 & 0 & 0 & 0 & 0 \\ 0 & 0 & 0 & 0 & 0 & 0 & 0 & 0 \end{bmatrix}.$$

Here r_{ij} denotes the coefficient of $x^i y^j$ for $0 \leqslant i, j \leqslant 7$ with the row–column index beginning at $(0, 0)$ at the left-top corner, i being the column index, and j the row index; e.g., the coefficient of $x^4 y^2$ is 46.

6.4 Complexity reduction

We already mentioned that the product of two m-variable polynomials, with degree $(n-1)$ in each variable, requires $\mu_D = n^{2m}$ multiplications. If FFT methods are used, as explained above, by appending n zeros to each sequence, the total complexity μ_C is as follows.

For two DFT and one IDFT we need

$$\frac{3m}{2}(2n)^m \log_2(2n)$$

operations and for point-wise product we need $(2n)^m$ multiplications or

$$\mu_C = \frac{3m}{2}(2n)^m \log_2(2n) + (2n)^m.$$

For $n = 4$, 8, 16, and $m = 2, 3, \ldots, 10$ we present in Table 6.1 the ratio μ_D/μ_C. It is seen that the FFT method would be more efficient for large m and n.

Table 6.1 Multivariable polynomial complexity reduction

m	μ_D/μ_C $n = 4$	μ_D/μ_C $n = 8$	μ_D/μ_C $n = 16$
2	0.4	1.2	4
3	0.6	3	22
4	0.8	10	130
5	1.35	33	800
6	2	110	
7	4	380	
8	7	1300	
9	12	4700	
10	21.0	16000	

6.5 EXAMPLE. Find the determinant of $A(x,y)$ using the Fourier evaluation–interpolation

$$A(x,y) = \begin{bmatrix} 2x & y+2 \\ x+2 & y+1 \end{bmatrix}.$$

We have $m = 2$, $n = 2$; we choose $M = 17$, $\omega = 16$.
  Evaluation:

$$A(1,1) = \begin{bmatrix} 2 & 3 \\ 3 & 2 \end{bmatrix}; \qquad A(16,1) = \begin{bmatrix} 15 & 3 \\ 1 & 2 \end{bmatrix}$$

$$A(1,16) = \begin{bmatrix} 2 & 1 \\ 3 & 0 \end{bmatrix}; \qquad A(16,16) = \begin{bmatrix} 15 & 1 \\ 1 & 0 \end{bmatrix}.$$

The evaluated determinants are $\det A(1,1) = 12$; $\det A(16,1) = 10$; $\det A(1,16) = 14$; $\det A(16,16) = 16$.
  Interpolation: We have

$$X_i^T = \begin{bmatrix} 1 & 1 \\ 1 & 16 \end{bmatrix}; \qquad Y_j^T = \begin{bmatrix} 1 & 1 \\ 1 & 16 \end{bmatrix};$$

thus interpolation of determinant is given by

$$\left[\begin{bmatrix} 1 & 1 \\ 1 & 16 \end{bmatrix}^{-1} \otimes \begin{bmatrix} 1 & 1 \\ 1 & 16 \end{bmatrix}^{-1}\right] \begin{bmatrix} 12 \\ 10 \\ 14 \\ 16 \end{bmatrix} = \begin{bmatrix} -4 \\ 0 \\ -2 \\ 1 \end{bmatrix}$$

or $\det A(x,y) = -4 - 2y + xy$.

EXERCISES III

1. Find ω such that $\omega^{18} \equiv 1 \bmod 37$ and $\omega^{9} \equiv 1 \bmod 37$.

2. Write a program to compute ω such that $\omega^n \equiv 1 \bmod p$, where p is a specified prime and n any integer such that $n \mid (p-1)$.

3. Find the product of the polynomials

$$\mathscr{P}(x) = 1 + 2x + 3x^2 + 4x^3$$
$$\mathscr{Q}(x) = 1 + 3x + x^2 + x^3$$

over $(\mathbb{I}_{37}, +, \cdot)$ using the NTT method.

4. Find the product of the polynomials

$$\mathscr{P}(x,y) = 1 + 2x + 3y + 4xy$$
$$\mathscr{Q}(x,y) = 2 + 3x + 4y + 8xy$$

using the NTT method. (The coefficients are integers.)

5. Find the inverse of the matrix using Fourier evaluation–interpolation

$$\begin{bmatrix} 2x & y+2 \\ x+1 & y+1 \end{bmatrix}.$$

6. Find the characteristic equation of A using the Fourier evaluation–interpolation method

$$A = \begin{bmatrix} 1-\lambda & 2 & 3 \\ 3 & 4-\lambda & 5 \\ 5 & 6 & 7-\lambda \end{bmatrix}.$$

(The elements of A are real.)

7. Multiply the following two matrix polynomials in $\mathbb{I}[x]$, using Fourier evaluation–interpolation

$$\begin{bmatrix} 3x+2 & 4x+6 \\ 2x+3 & 6x+7 \end{bmatrix} \begin{bmatrix} x+1 & x+5 \\ x+3 & x+6 \end{bmatrix}.$$

8. Using Fourier evaluation–interpolation find the product of

$$\begin{bmatrix} 1 & x \\ 1 & y \end{bmatrix} \begin{bmatrix} x & y \\ y & x \end{bmatrix}.$$

9. Use Fourier evaluation–interpolation to compute

$$\begin{bmatrix} 1 & x \\ 1 & y \end{bmatrix}^2.$$

10. Invert the following matrix using Fourier evaluation–interpolation

$$A = \begin{bmatrix} x & y \\ y & z \end{bmatrix}.$$

CHAPTER IV
Polynomial Hensel Codes

1 Introduction

Hensel's p-adic analysis (Koblitz [1977]) plays a central role in mathematical research serving as a bridge between classical analysis and algebra. This chapter and the following chapter are concerned with the application of p-adic technique as an artifice for performing symbolic computations in linear algebra. We introduce in this chapter, three topics that are of primary interest:

(i)   The theory and construction of Hensel Codes for arithmetic of rational functions (single variable) over a finite field, analogous to the arithmetic of rational numbers using Hensel Codes (finite segment p-adic numbers) (Gregory and Krishnamurthy [1984]).

(ii) The application of Hensel Codes for the solution of linear algebraic problems.

(iii) The use of Hensel's lemma (Koblitz [1977], as a powerful tool for devising fast, parallel, exact, and symbolic algorithms for the inversion of matrices whose elements are Hensel Codes.

In order to explain the Hensel Code and related concepts, we need to introduce briefly the notion of Hensel fields and isomorphic algebras.

2 Hensel Fields

We define a valued field as a triple $(F, \mathbb{I}, \text{ord})$ where F is a field and $\mathbb{I}$ is an ordered abelian group $(\mathbb{I} \neq 0)$; $\mathbb{I}$ is called the value group and ord $F^* \to \mathbb{I}$ is a surjective map $(F^* = F - \{0\}$ is the multiplicative group of nonzero elements of F) satisfying:

$$\text{for } a, b \in F^*$$

$$\text{ord}(ab) = \text{ord } a + \text{ord } b$$

$$\text{ord}(a + b) \geq \min(\text{ord } a, \text{ord } b)$$

(see Section 3.2, Chapter I).

Two typical examples of valued fields are $\mathbb{Q}_p$ the p-adic field (Koblitz [1977], Gregory and Krishnamurthy [1984]) and the field of formal power series over F (Section 3.2.9, Chapter I). They are called Hensel fields or complete valued fields satisfying the following properties (Cherlin [1976], Mahler [1981]).

Let $(F, \mathbb{I}, \text{ord})$ be a valued field, whose value group $\mathbb{I}$ is the group of integers under addition. Let $\{a_i\}$ be the sequence of elements of F and a be an element of F.

(i) a_i converges to a, if and only if, $\text{ord}(a_i - a) \to \infty$;
(ii) a_i is a Cauchy sequence, if and only if,

$$\text{ord}(a_i - a_j) \to \infty \quad \text{as } i, j \to \infty;$$

(iii) F is complete, if and only if, every Cauchy sequence in F converge to some limit.

Note: A sequence $\{a_i\}$ is Cauchy, if and only if, $\text{ord}(a_i - a_{i+1}) \to \infty$, as i increases. This follows from the definition of a valued field which satisfies

$$\text{ord}(a + b) \geq \min(\text{ord } a, \text{ord } b).$$

2.1 Hensel's lemma

One very important algebraic theorem in complete valued fields is the Hensel's lemma.

Let $(F, \mathbb{I}, \text{ord})$ be a complete valued field in the integers. Let $f(x) \in R[x]$ be a polynomial with coefficients in the valuation ring $R(R = \{a \in F : \text{ord } a \geq 0\})$ and let $\bar{f}(x) \in \bar{F}[x]$ be the result of reducing the coefficients of f modulo the valuation ideal M. Suppose that $\bar{f}$ has a simple root, α in $\bar{F}$ (or explicitly, $\bar{f}(\alpha) = 0$ and $\bar{f}'(\alpha) \neq 0$); then f has a root a in the valuation ring such that $\bar{a} = \alpha$.

Hensel's lemma applies to both the p-adic field and the formal power series field (Section 3, Chapter I).

2.2 Bibliographic remarks

The metamathematical connection between the p-adic fields $\mathbb{Q}_p$ of the algebraic number theory and the formal power series fields $F_p((x))$ was established by Ax and Kochen [1965] when they formulated their transfer principle: an elementary statement (a statement formed by using the logical connectors *and, or, not, if, then, if and only if, equality,* symbols of variables $x_1, x_2, \ldots$, quantifiers $\exists$, $\forall$, and the relational symbols) about valued fields is valid in the p-adic field $\mathbb{Q}_p$, if and only if, it is valid in $F_p((x))$ for all but a finite number of primes p.

Paul Cohen [1969] is concerned with how to decide by means of a recursive procedure, elementary statements about real and p-adic fields. Both these papers are of fundamental importance in Computer Science.

3 Isomorphic Algebras

One of the most elegant and useful concepts of universal algebra is the concept of morphism—which is a structure preserving map from one algebraic system to another. We will briefly explain in this section how this concept is useful in computation.

3.1 Similar algebras

We first look at the notion of "similar algebras" Cohn [1965]. Two algebras (A,P) and (A',P') are similar, if there is a bijection from P to P' with the property that the corresponding operations $p \in P$ and $p' \in P'$, have the same arity (or corresponding nullary, unary, binary, ..., operations).

3.2 Morphism

Let (A,P) and (A',P') be similar algebras. A function

$$\phi : A \to A'$$

is said to be a morphism (or homomorphism) from (A,P) to (A',P'), if for any $p \in P$ *and corresponding* $p' \in P'$, and for any $a_1, a_2, \ldots, a_n \in A$,

$$(3.1) \qquad \phi(p(a_1, a_2, \ldots, a_n)) = p'(\phi(a_1), \ldots, \phi(a_n)).$$

In other words, the morphism ϕ preserves corresponding operations of A and A'. That is, the same result is obtained whether

(i) an operation is performed in A and the result is mapped into A' through ϕ (using (3.1)); or

(ii) the arguments from A are first mapped into A' (through ϕ) and then the corresponding operation of A' is performed as in the right hand side of (3.1).

3.2.1 *Classification of morphisms.* If $\phi : A \to A'$ is a morphism then we call ϕ

(i)   an ISOMORPHISM if ϕ is bijective;
(ii) an EPIMORPHISM if ϕ is surjective;
(iii) a MONOMORPHISM if ϕ is injective.

Thus A and A' are called "isomorphic" when ϕ is bijective. In such a case, A' can be considered as a copy of A which results from ϕ assigning different names (semantic interpretation) to the elements of A. Therefore, any algebraic law satisfied by A must also be satisfied by A'. If A is a model exemplifying some axiom system, then so is A' (Cohn [1965]).

3.2.2 EXAMPLE. The following familiar example illustrates the practical utilization of the concept of Isomorphism.
    Consider the map

$$\phi : x \to \log x$$

from $(\mathbb{R}^+, \cdot)$ the set of positive reals under multiplication, to $(\mathbb{R}, +)$ the set of all reals under addition. Then

$$\phi(x \cdot y) = \log x \cdot y = \log x + \log y$$
$$= \phi(x) + \phi(y)$$

so that ϕ is a morphism; also ϕ is bijective with the inverse given by the exponential function $x \to \exp x$; thus ϕ is an Isomorphism.

This mapping enables us to carry out the multiplication operation of two positive reals by the three step procedure:

(i)   a forward mapping—taking logs of two operands;
(ii) calculus—adding the logs;
(iii) an inverse mapping—exponentiating.

Although two isomorphic algebras have identical properties their computational properties may be quite different. In the above example, for instance, the complexity of operations may widely differ in their implementation and performance.

3.3 Bibliographic remarks

In classical algebra, isomorphism implies that any algebraic property which A (a ring or a field) possesses is shared by A' (a ring or a field). But this is no longer true of non-purely-algebraic properties—such as effective computability: in an arbitrary isomorphism $A \leftrightarrow A'$, there may be no algorithm for finding the element $a \in A$ given $a' \in A'$ and vice-versa. Hence Shepperdson and Frohlich [1956] define the new concept of "explicit or constructive isomorphism". An isomorphism $\phi : A \to A'$ is said to be an explicit isomorphism from (A,P) to (A',P'), if for any $p \in P$ and corresponding $p' \in P'$, and for $a_1, a_2, \ldots, a_n \in A$, we have an algorithm (or effective procedure) to compute $\phi(p(a_1, \ldots, a_n))$ and also an algorithm to decide whether the equality (3.1)

$$\phi(p(a_1, \ldots, a_n)) = p'(\phi(a_1), \ldots, \phi(a_n))$$

holds.

Shepperdson and Frohlich [1956] also call a ring R (field F) explicit, if there are algorithms to compute the sum, product, subtraction (division), and algorithms for finding unit element, null element, and testing for equality of elements. For further details on effective mathematics, see: Robinson [1976], Davenport [1981], Buchberger, Collins and Loos [1983], Tourlakis [1984], Kfoury, Moll, and Arbib [1982], Aberth [1980].

3.4 Hensel Codes for rational numbers

In Gregory and Krishnamurthy [1984] it is explained how the algebra of rational numbers can be realized by using the "explicit isomorphism" between the algebra of rational numbers and the ring of integers modulo an integer (or the finite segment Hensel's p-adic representation). The practical implementation consists of the following three step process:

(i) Forward mapping of the set of order-N Farey rationals $\mathbb{F}_N = \{a/b \in \hat{\mathbb{Q}} : \gcd(a,b) = 1$ and $0 \le |a| \le N, 0 < |b| \le N\}$, where $\hat{\mathbb{Q}} = \{a/b : \gcd(b,p) = 1\}$, into $\mathbb{I}_m$ where $m = p^r$; $\mathbb{I}_m$ is called the Hensel Code or a finite segment p-adic representation of $\mathbb{F}_N$;

(ii) carry out the arithmetic operations in the ring $(\mathbb{I}_m, +, \cdot)$ free of representation errors; and

(iii) map the integer (or Hensel Code) results back into $\mathbb{F}_N$.

It was shown in Gregory and Krishnamurthy [1984] that if

$$N \le \sqrt{\frac{p^r - 1}{2}}$$

then the mapping

$$|\cdot|_m : \mathbb{F}_N \to \hat{\mathbb{I}}_m,$$

where $\hat{\mathbb{I}}_m = \{|ab^{-1}|_m : a/b \in \mathbb{F}_N\}$ is one-to-one and onto and thus has an inverse (note b^{-1} exists since $m = p^r$ and $\gcd(b,p) = 1$); or the mapping $|\cdot|_m$ is an explicit isomorphism by virtue of the existence of the algorithms for forward mapping $\mathbb{F}_N \to \hat{\mathbb{I}}_m$, the arithmetic in $(\mathbb{I}_m, +, \times)$, and the inverse mapping from $\mathbb{I}_m \to \mathbb{F}_N$ through the extended Euclidean algorithm (Chapter I).

Our aim in the next section is to extend by analogy, all the concepts regarding the construction of Hensel Codes and arithmetic operations, to the rational polynomial functions over the ring of integers, through the use of formal power series fields (Section 2).

4 Hensel Codes for Rational Polynomials

We are now ready to use the algebraic concepts outlined earlier, for the construction of Hensel Codes for rational polynomials. For this purpose we need to introduce some notational convention and new definitions.

4.1 Definitions and notation

We define the following sets:

(a) $\mathbb{I}[x] = \{a(x) = \sum_{i=0}^{m} a_i x^i : a_i \in \mathbb{I}$, and m, a nonnegative integer$\}$
(b) $\mathbb{I}_p[x] = \{a(x) = \sum_{i=0}^{m} a_i x^i : a_i \in \mathbb{I}_p$, and m, a nonnegative integer$\}$
(c) $\mathbb{I}_p[[x]] = \{a(x) = \sum_{i=0}^{\infty} a_i x^i : a_i \in \mathbb{I}_p\}$
(d) $F(x) = \{a(x)/b(x) : a(x), b(x) \in \mathbb{I}[x], b(x) \neq 0\}$
(e) $F_p(x) = \{a(x)/b(x) : a(x), b(x) \in \mathbb{I}_p[x], b(x) \neq 0\}$
(f) $\hat{F}_p(x) = \{a(x)/b(x) : a(x), b(x) \in \mathbb{I}_p[x], \gcd(b(x), x) = 1, b(x) \neq 0\}$
(g) $P(L/M,N,x) = \{a(x)/b(x) \in F(x) : \gcd(a(x), b(x)) = 1,\ 0 \leq |a_i|,\ |b_i| \leq N,$
  $0 \leq \deg a(x) \leq L, 0 \leq \deg b(x) \leq M\}$.

We call $P(L/M,N,x)$ order-N, degree (L/M) Padé rational polynomial. As seen from (g), this is a rational polynomial whose numerator polynomial and denominator polynomial have integral coefficients with maximal coefficient size not exceeding N, and the numerator is of degree at most L and the denominator is of degree at most M. (Baker [1975], Baker and Graves-Morris [1981].)

(h) $P(L/M, F_p(x)) = \{a(x)/b(x) \in F_p(x) : \gcd(a(x), b(x)) = 1,\ 0 \leq \deg a(x) \leq$
  $L, 0 \leq \deg b(x) \leq M\}$.

4.1.1 **Remark.** The Padé rational polynomial $P(L/M,N,x)$ coincides with the order-N Farey rational (Gregory and Krishnamurthy [1984] for $L = M = 0$).

4.2 Forward mapping

The forward mapping is a succession of two mappings:

(i) $P(R-1/R-1, N, x) \rightarrow P(R-1/R-1, \hat{F}_p(x))$.

Here we map the rational polynomial (with integer coefficients) whose numerator and denominator degrees are at most $(R-1)$ and maximum coefficient size N to the rational polynomial $\hat{F}_p(x)$ using the following rules, after factoring out purely powers of x, if any, in $a(x)$ and $b(x)$.

(4.1)   Choose $p \geqslant 2RN^2 + 1$

(see Section 6.2, Chapter III) and set

$$
\begin{aligned}
a_i &\to a_i \quad \text{if} \quad a_i \geqslant 0 \\
a_i &\to (p + a_i) \quad \text{if} \quad a_i < 0
\end{aligned}
$$

(4.2)

(see Section 4.5, Chapter I). Similarly

$$
\begin{aligned}
b_i &\to b_i \quad \text{if} \quad b_i \geqslant 0 \\
b_i &\to (p + b_i) \quad \text{if} \quad b_i < 0.
\end{aligned}
$$

(4.3)

(ii) $P(R - 1/R - 1, \hat{F}_p(x)) \to \hat{\mathbb{I}}_p[[x]] \to \mathbb{I}_p[x]$.

Here we first map the Padé rational polynomial over a finite field to its formal power series in $\hat{\mathbb{I}}_p[[x]]$ by finding $[a(x) \cdot b(x)^{-1}]$. This infinite series is truncated and mapped to a subset $\hat{\mathbb{I}}_p[x]$ of $\mathbb{I}_p[x]$ by the rule

$$\hat{\mathbb{I}}_p[x] = \{[a(x) \cdot b(x)^{-1}] \bmod x^{2R-1} : a(x)/b(x) \in P(R - 1/R - 1, \hat{F}_p(x))\}.$$

In other words, $\hat{\mathbb{I}}_p[x]$ denotes the set of images of the Padé rationals $P(R - 1/R - 1, N, x)$.

The following theorems ensure that the mappings described above are one-to-one and onto and thus have an inverse.

4.3 Theorem. *Let $\alpha(x) = a(x)/b(x)$ and $\beta(x) = c(x)/d(x)$ be the elements of $P(R - 1/R - 1, N, x)$ and let these be mapped to $P(R - 1/R - 1, \hat{F}_p(x))$ using the rules:*

$$
\begin{aligned}
a_i &\to a_i, \quad \text{if } a_i \geqslant 0 \\
a_i &\to p + a_i, \quad \text{otherwise};
\end{aligned}
$$

and

$$
\begin{aligned}
b_i &\to b_i, \quad \text{if } b_i \geqslant 0 \\
b_i &\to p + b_i, \quad \text{otherwise};
\end{aligned}
$$

where $p \geqslant 2N^2R + 1$.

Then $\alpha(x)$ and $\beta(x)$ map to the same element in $P(R - 1/R - 1, \hat{F}_p(x))$ if and only if

(4.4) $a(x)d(x) \equiv b(x)c(x) \bmod p.$

PROOF. Since $b(x)$ and $d(x)$ are not zero (by definition) and $F = (\mathbb{I}_p, +, \cdot)$ is a field, we can assume that $b(x)^{-1}$ and $d(x)^{-1}$ exist as a formal power series in $F[[x]]$. Thus multiplying both sides of (4.4) by $b(x)^{-1}d(x)^{-1}$ we get

$$a(x)b^{-1}(x) \equiv c(x)d^{-1}(x) \bmod p$$

or $|\alpha(x)|_p \equiv |\beta(x)|_p$. The converse is proved similarly. $\square$

4.4 Theorem. Let $\alpha(x) = a(x) \cdot b^{-1}(x)$ be a formal power series over $F = (\mathbb{Q}_p, +, \cdot)$. Let $\alpha^T(x) \equiv \alpha(x) \bmod x^{2R-1}$.

Then $\alpha^T(x)$ is unique for any element in $P(R - 1/R - 1, \hat{F}_p(x))$.

PROOF. If we assume that corresponding to α^T there exist two elements $f(x)/g(x), \gcd(f(x), g(x)) = 1$, and $h(x)/m(x), \gcd(h(x), m(x)) = 1$ then we must have

(4.5) $\mathrm{ord}(f(x)/g(x) - h(x)/m(x)) = (2R - 1)$

since both approximate the same series. Multiplying (4.5) by $g(x)m(x)$

(4.6) $\mathrm{ord}(f(x)m(x) - h(x)g(x)) = (2R - 1)$.

But the left hand side of (4.6) is a polynomial of degree at most $(2R - 2)$ (by Theorem 3.2.3 of Chapter I) and thus is identically zero. Since neither $h(x)$ nor $m(x)$ is identically zero and by assumption both $f(x)$ and $g(x)$, and $h(x)$ and $m(x)$ are relatively prime, the two elements are the same. $\square$

4.5 Theorem (Uniqueness of Coding). *Let N be the largest integer satisfying the inequality*

(4.7) $$N \leq \sqrt{(p-1)/2R}.$$

Then in order that $a(x)/b(x)$ belonging to $P(R - 1/R - 1, N, x)$ be represented uniquely by $\alpha^T(x)$ in $\hat{\mathbb{Q}}_p[x]$ it is sufficient that

(i) deg $a(x)$ and deg $b(x)$ are both at most $(R - 1)$; and
(ii) the maximal absolute value of the coefficients $|a_i|, |b_i| \leq N$.

PROOF. (i) We already proved in Theorem 4.4 that a truncated series α^T can have a unique correspondence with a rational if the left hand of (4.6) has a degree at most $(2R - 2)$ since the truncation is done mod x^{2R-1}. As this requirement is met with, we can have a unique $\alpha^T(x)$.

(ii) Let $\alpha(x) = a(x)/b(x), \gcd(a(x), b(x)) = 1$, and $\beta(x) = c(x)/d(x)$, $\gcd(c(x), d(x)) = 1$ be two different rational functions belonging to $P(R - 1/R - 1, N, x)$ mapping to the same polynomial in $\hat{\mathbb{Q}}_p(x)$. Then by Theorem 4.3

$$a(x)d(x) \equiv b(x)c(x) \bmod p$$

(4.8)   or

$$a(x)d(x) - b(x)c(x) \equiv 0 \bmod p.$$

If we assume that the modulus of the coefficients in $a(x), b(x), c(x)$ and $d(x)$ are all less than N, then the polynomial

$$a(x)d(x) - b(x)c(x) = f(x)$$

has coefficients lying in the range $0 \leq f_i \leq 2N^2R$.

Hence a choice $p \geq 2N^2R + 1$ ensures that (4.8) is an equality rather than a congruence or

$$a(x)d(x) - b(x)d(x) = 0$$
$$a(x)/b(x) = c(x)/d(x).$$

Hence $\alpha(x)$ and $\beta(x)$ are the same.   □

4.6 Hensel Codes

The Hensel Code for a rational polynomial $a(x)/b(x)$ is simply the truncated infinite series expansion $[a(x) \cdot b(x)^{-1}] \bmod x^{2R-1}$ which belongs to $\hat{\mathbb{I}}_p[x]$.

Note that in our definition of $\hat{F}_p(x)$ in Section 4.1(f) we excluded the possibility of $b(x)$, the denominator polynomial being purely a positive power of x or containing a factor which is a positive power of x. This is because of our requirement to obtain a formal power series expansion in which negative powers are not permissible. However, from the point of view of practical computation, such factors can be kept separately as exponents, as we did earlier for the powers of p, in obtaining the Hensel Codes for rational numbers, (or we treat such a case as an extended power series $F\langle x\rangle$; Section 3.2.9, Chapter I). Thus while mapping from $P(R - 1/R - 1, N, x)$ to $P(R - 1, R - 1, \hat{F}_p(x))$ we remove powers of x in the numerator and denominator and keep count. In other words, the given rational polynomial is expressed as

$$\alpha(x) = (a(x)/b(x)) \cdot x^n,$$

where $a(x)/b(x) \in \hat{F}_p(x)$ and n is a positive or negative integer. The Hensel Code of $\alpha(x) \in P(R - 1/R - 1, N, x)$ is denoted by $H(p, r, \alpha(x))$, where $r = 2R - 1$. This consists of an ordered pair : $(m_\alpha; e_\alpha)$, where $m_\alpha =$ mantissa that represents the truncated infinite series expansion $[a(x) \cdot b(x)^{-1}] \bmod x^{2R-1} \in \mathbb{I}_p[x]$ and $e_\alpha =$ exponent of value n. For convenience, it is sufficient to record only the coefficients in m_α (in increasing degree) separated by a delimiter (,) and e_α as a mere positive or negative integer.

4.6.1 EXAMPLES. (i) Let $\alpha(x) = 4 + 2x + 3x^2 + 6x^3 + \cdots \in \mathbb{I}_7[x]$. Then

$$H(7, 4, \alpha(x)) = (4, 2, 3, 6; 0).$$

(ii) Let

$$\alpha(x) = \frac{x^2 + x^3}{1 + x^3} = \frac{1 + x}{1 + x^3} \cdot x^2 \in F_3(x)$$

$$\begin{aligned}
H(3, 6, \alpha(x)) &= x^2(1 + x) \cdot (1 + x^3)^{-1} \\
&= x^2 \cdot (1 + x)(1 + 0 \cdot x + 0 \cdot x^2 + 2x^3 + 0 \cdot x^4 + 0 \cdot x^5) \\
&= x^2(1 + 1 \cdot x + 0 \cdot x^2 + 2x^3 + 2x^4 + 0 \cdot x^5) \\
&= (1, 1, 0, 2, 2, 0; 2).
\end{aligned}$$

4.6.2 EXAMPLES (Unique Hensel Codes). (i) Let $\alpha(x) = (2 + x/1 - x) \in P(1/1, 2, x)$; we need to choose $p \geqslant 2.2.4 + 1 = 17$ since $R - 1 = 1$ or $R = 2$ and $N = 2$. Therefore we choose $p = 17$ and first map $(2 + x)/(1 - x)$ to $\hat{\mathbb{I}}_{17}[x]$. We thus obtain

$$\alpha(x) = (2 + x)/(1 + 16x).$$

Then we convert $\alpha(x)$ into a truncated formal power series over $\mathbb{I}_{17}$ thus:

$$(1 + 16x)^{-1} \bmod x^3 = 1 + x + x^2.$$

Thus

$$(2 + x)(1 + x + x^2) \bmod x^3 = 2 + 3x + 3x^2 \in \mathbb{I}_{17}[x]$$

or

$$H(17, 3, 2 + x/1 - x) = (2, 3, 3; 0).$$

(ii) Let

$$\alpha(x) = 2 + x \in P(2/2, 10, x).$$

We need to choose a prime $p \geqslant 2RN^2 + 1 = 601$. We choose $p = 601$. Then

$$H(601, 3, 2 + x) = (2, 1, 0; 0).$$

4.6.3 **Remark.** It is clear from the above examples that by a proper choice of R, N and hence p, the Hensel Code of $\alpha(x) \in F(x)$ can be used in an exactly analogous manner as that for a rational number for performing exact arithmetic with rational polynomials, provided we carry out the forward and inverse mapping.

4.6.4 **Remark.** There is a close relationship between Goppa Codes used for error correction, Padé approximants, and the Hensel Codes; all these share the property of having in common the extended Euclidean algorithm for decoding (McEliece [1977], McEliece and Shearer [1978], Brent, Gustavson and Yun [1980], Yun [1983]).

4.7 Forward mapping

The forward mapping is the problem of conversion of $a(x)/b(x) = \alpha(x) \in P(R - 1/R - 1, N, x)$ to its Hensel Code $H(p, 2R - 1, \alpha(x))$. As already illustrated in examples, we can compute $b^{-1}(x)$ and multiply by $a(x) \bmod x^{2R-1}$. Alternatively, one can use the analogue of the Algorithm 2.8 of Chapter I, to obtain $a(x) \cdot b^{-1}(x) \bmod x^{2R-1}$, by choosing

$$a = x^{2R-1}; \qquad b = 0; \qquad c = b(x); \qquad d = a(x)$$

(see also Algorithm 3.6.4 of Chapter I).

4.8 Inverse mapping

The inverse mapping is the problem of conversion of the Hensel Code $H(p, 2R - 1, \alpha(x))$ back into $\alpha(x) \in P(R - 1/R - 1, N, x)$. This consists of two mappings:

$$M_1 : H(p, 2R - 1, \alpha(x)) \rightarrow P(R - 1/R - 1, \hat{F}_p(x));$$
$$M_2 : P(R - 1/R - 1, \hat{F}_p(x)) \rightarrow P(R - 1/R - 1, N, x), \quad \text{where } N$$
$$\leqslant \sqrt{(p - 1)/2R}.$$

4.8.1 *Mapping M_1.* This can be achieved by any one of the following methods (see Sections 6.2, 6.3, and 6.4):

(a) Padé method. Here we solve a linear system of equations with a coefficient matrix of size $(2R - 1) \times (2R - 1)$, which is lower triangular. This matrix is known as Padé matrix. This results in a rational polynomial $\alpha(x)$ whose numerator and denominator belong to $I_p[x]$ and denominator has its constant term $b_0' = 1$ or

$$\alpha(x) = \frac{a_0' + a_1'x + \cdots + a_{R-1}'x^{R-1}}{b_0' + b_1'x + \cdots + b_{R-1}'x^{R-1}},$$

where a_i' $(i = 0, 1, \ldots, R - 1)$ and b_i' $(i = 1, 2, \ldots, R - 1)$ belong to I_p and each represent a Padé rational P_N whose numerator and denominator are $\leqslant N$, where $N = \sqrt{p - 1/2R}$ (Section 6.2).

(b) Euclidean method. This algorithm is analogous to that used for inverse mapping of modulo m integers back into Farey rationals F_N described in Gregory and Krishnamurthy [1984] (see also Section 4.8.2.) Here we use the seed matrix

$$\begin{bmatrix} x^{2R-1} & 0 \\ H(p, 2R - 1, \alpha(x)) & 1 \end{bmatrix}$$

and obtain a rational polynomial $\in F_p(x)$ where the numerator polynomial $a(x)$ and denominator polynomial $b(x)$ have coefficients in I_p, each coefficient representing a Padé rational whose numerator and denominator are $\leqslant N$, where $N = \sqrt{(p - 1)/2R}$ (see Section 6.3).

(c) Common denominator method. The common denominator method (Gregory and Krishnamurthy [1984]; Section 6.40; Chapter I) is based on the following simple principle: As every rational polynomial $a(x)/b(x)$ is represented in the form $a(x)b(x)^{-1} \bmod x^{2R-1}$, it is possible to determine $a(x)$ as well as $b(x)$, if some common multiple of all the denominators involved in a given algorithm is known. Since any algorithm consists of a predetermined sequence of arithmetic operations, it is possible to derive the expression for this common multiple to facilitate the conversion (see Section 6.4).

4.8.2 *Mapping* M_2. Here we convert each $a_i, b_i \in \mathbb{I}_p$ into a Padé rational

$$P_N = \left\{ \frac{a}{b} : 0 \leqslant |a| \leqslant N, 0 < |b| \leqslant N \quad \text{and} \quad N \leqslant \sqrt{(p-1)/2R} \right\}$$

by the inverse mapping algorithm (based on Euclidean algorithm, Section 2.8, Chapter I). The common denominator between numerator and denominator polynomials can then be cancelled to obtain $P(R-1/R-1, N, x)$.

We use the seed matrix

$$\begin{bmatrix} p & 0 \\ a_i & 1 \end{bmatrix}$$

in Algorithm 2.8 of Chapter I and terminate when A_i and B_i are less than or equal to N; then a_i corresponds to the Padé rational A_i/B_i.

4.8.3 EXAMPLE. Convert $403 \in \mathbb{I}_{601}$ into P_N where $p = 601$, $R = 3$,

$$N \leqslant \sqrt{(601-1)/2 \cdot 3} = 10.$$

We start with the seed matrix

$$\begin{bmatrix} 601 & 0 \\ 403 & 1 \end{bmatrix}$$

and obtain the following table:

	601	0
	403	1
1	198	−1
2	7	3
28	2	−85

Thus 403 maps to the rational 7/3.

5 Arithmetic of Hensel Codes

We now describe the arithmetic operations on rational polynomial functions using polynomial Hensel Codes.

5.1 Complementation/negation

Given $H(p, r, \alpha(x))$, the code of $H(p, r, -\alpha(x))$ is obtained by a digitwise radix complementation modulo p. In other words, if

$$H(p, r, \alpha(x)) = (\alpha_0, \alpha_1, \ldots, \alpha_{r-1}; n)$$

then the complement $H(p, r, \bar{\alpha}(x))$ is obtained by the operations

$$\begin{cases} (p - \alpha_i) \rightarrow \alpha_i, & \text{if } \alpha_i \neq 0; \\ 0 \rightarrow \alpha_i, & \text{otherwise.} \end{cases}$$

With n remaining the same we have then

$$H(p, r, -\alpha(x)) = H(p, r, \bar{\alpha}(x)).$$

5.2 Addition/subtraction

Let

$$H(p, r, \alpha(x)) = (\alpha_0, \ldots, \alpha_{r-1}; e_\alpha)$$
$$H(p, r, \beta(x)) = (\beta_0, \ldots, \beta_{r-1}; e_\beta).$$

The algorithm for addition aligns the corresponding degree terms of the mantissa, retaining the lower exponent and finds the sum digit s_i using

$$s_i = (\alpha_i + \beta_i) \bmod p \quad \text{for} \quad i = 0, 1, 2, \ldots, r-1.$$

After the addition is complete, the result is expressed in (m_α, e_α) form.

Subtraction is realized as complemented addition.

5.2.1 EXAMPLES

$$H(5, 6, \alpha(x)) = (4, 3, 3, 3, 3, 3; 0)$$
$$H(5, 6, \beta(x)) = (2, 3, 1, 3, 1, 3; -2)$$
$$H(5, 6, \alpha(x) + \beta(x)) = (2, 3, 0, 1, 4, 1; -2)$$
$$H(5, 6, \alpha(x) - \beta(x)) = (3, 2, 3, 0, 2, 0; -2).$$

5.3 Multiplication

This is exactly similar to the polynomial multiplication modulo p except that the product is developed only up to r digits; higher order digits are discarded (modulo x^r).

The multiplication algorithm consists in forming the cross product of the digits of mantissa:

$$P_{ij} = \beta_i \alpha_j \bmod p$$

for $i = 0, 1, \ldots, (r-1)$, and $j = 0, 1, 2, \ldots, (r-1)$ and then forming the partial products P_i and the final product P by shifts and additions as specified by the following recursions for each i $(0 \leqslant i \leqslant r-1)$

$$P_i = \sum_{j=0}^{r-1} P_{ij} \Delta(j)$$

$$P = \sum_{i=0}^{r-1} P_i \Delta(i) \bmod p,$$

where $\Delta(x)$ denotes a right shift by x positions. The exponent of the product is $(e_\alpha + e_\beta)$.

5.3.1 EXAMPLE. $p = 5$: $r = 4$: $H(5, r, \alpha(x)) = (4, 3, 3, 3; 0)$; $H(5, 4, \beta(x)) = (2, 3, 1, 3; 0)$

i, j	0	1	2	3
α_i	4	3	3	3
β_i	2	3	1	3
P_0	3	1	1	1
P_1		2	4	4
P_2			4	3
P_3				2
P	3	3	4	0

Thus $H(3, 4, \alpha(x) \cdot \beta(x)) = (3, 3, 4, 0; 0)$.

5.4 Multiplicative inverse and division

As in the case of rational numbers (Gregory and Krishnamurthy [1984]) here we can have three different algorithms. One of them is direct involving the recursive solution of congruences with respect to p. The other methods are based on obtaining the reciprocal either by Newton's iteration or by the use of Euclidean gcd algorithm. We will describe the latter methods in the following subsections.

5.4.1 *Direct division.* Let $\beta = \beta_0, \beta_1 \cdots \beta_{r-1} (\beta_0 \neq 0)$ and $\beta^{-1} = q_0, q_1 \cdots q_{r-1}$, such that $\beta \cdot \beta^{-1} = 1, 0, \ldots, 0 \bmod x^r$. Then the q_i's are obtained by solving for q_i in

$$\beta \sum_{i=0}^{r-1} q_i p^i \equiv 1 \bmod p.$$

Thus starting with $q_0 = \beta_0^{-1} \bmod p$ and assuming q_{k-1} is known, each $q_k (1 \leq k \leq r - 1)$ can be computed uniquely solving

$$\sum_{j=0}^{k} q_j \beta_{k-j} \equiv 0 \bmod p.$$

Based on this principle the division can be carried out by the following algorithm.

Let R_0 = zeroth partial remainder or numerator $\alpha(x)$ ($= 1$ for finding $\beta(x)^{-1}$)

$$R_{ii} = i\text{th positional digit of } R_i.$$

Then $q_i \equiv R_{ii} \cdot \beta_0^{-1} \bmod p$ for $i = 0, 1, \ldots, (r-1)$

$$R_{i+1} = R_i - q_i\beta(x)\Delta(i) \quad \text{(modulo } p, \text{ digitwise)}$$
$$= R_i + q_i \cdot \bar{\beta}(x)\Delta(i),$$

where $\Delta(i) = \text{right shift by } i \text{ digits and } \bar{\beta}(x) = \text{complement of } \beta(x)$.

5.4.2 **Remarks.** This algorithm can be applied to find β^{-1} if $R_{00} = 1$ and all other digits of R_0 are set to zero.

If $\beta_i = 0$ shift the divisor keeping count i until the first nonzero digit is encountered and suitably adjust the exponent. If i is the number of shifts made and e_α and e_β are the exponents of numerator and denominator then the exponent of the quotient is $e_q = e_\alpha - e_\beta - i$.

5.4.3 EXAMPLE. Let $H(7, 4, \alpha(x)) = (2, 3, 4, 5; 0)$ and $H(7, 4, \beta(x)) = (2, 1, 4, 2; 0)$. Then

$$H(7, 4, \bar{\beta}(x)) = (5, 6, 3, 5; 0)$$
$$\beta_0^{-1} \bmod 7 = 4.$$

The following table shows the computation:

i	0	1	2	3	$q_i = R_{ii} \cdot \beta_0^{-1} \bmod p$
R_0	2	3	4	5	$q_0 = 2 \cdot 4 \bmod 7 = 1$
$q_0\bar{\beta}\Delta(0)$	5	6	3	5	
R_1	0	2	0	3	$q_1 = 2 \cdot 4 \bmod 7 = 1$
$q_1\bar{\beta}\Delta(1)$		5	6	3	
R_2		0	6	6	$q_2 = 6 \cdot 4 \bmod 7 = 3$
$q_2\bar{\beta}\Delta(2)$			1	4	
R_3			0	3	$q_3 = 3 \cdot 4 \bmod 7 = 5$
$q_3\bar{\beta}\Delta(3)$				4	
R_4				0	

Thus $H(7, 4, \alpha(x) \cdot \beta(x)^{-1}) = (1, 1, 3, 5; 0)$.

5.5 Newton–Hensel–Zassenhaus (NHZ) iterative method

The Newton's method for finding a root of some differentiable function $f(x)$ is based on the first order Taylor-series expansion of f about some point x_i

$$f(x) = f(x_i) + f'(x_i) \cdot (x - x_i).$$

Assuming x is, or is sufficiently near, a root of f so that $f(x) = 0$ we obtain

(5.1) $x = x_i - f(x_i)/f'(x_i)$

as the Newton's iterative formula.

If we are given the function $f(x) = (1/x) - a$, the Newton's iteration can be used to compute the reciprocal of a to any desired accuracy over the real field, using (5.1) thus:

(5.2)
$$x_{i+1} = x_i - \left(\frac{1}{x_i} - a\right) / \left(\frac{-1}{x_i^2}\right)$$
$$= 2x_i - ax_i^2 = x_i(2 - ax_i), \qquad i = 0, 1, 2, \ldots.$$

The sequence of approximation produced by (5.2) converges to the reciprocal of a, quadratically; i.e., if x_0 contains at least one significant digit of accuracy, then the number of significant digits double during each step of the iteration.

In Gregory and Krishnamurthy [1984], it is explained in Chapter II, how the Newton's method can be used to find the reciprocal of p-adic numbers $\mathbb{Q}_p$. This is a consequence of Hensel's lemma (Section 2.1). Since both $\mathbb{Q}_p$ and the formal power series field are Hensel fields, we can apply Hensel's lemma to find the reciprocal of a formal power series. Our aim in this section is to extend the Newton's method to polynomial Hensel Codes. For this purpose, we will restate Hensel's lemma (Section 2.1) and its variation proposed by Zassenhaus [1969] (see also Yun [1976], Lipson [1981]).

5.5.1 **Hensel's Lemma.** *Let p be a prime in $\mathbb{I}$ and $f(x)$ be a given polynomial in $\mathbb{I}[x]$. Let $G_1(x)$ and $H_1(x)$ be two relatively prime polynomials in $\mathbb{I}_p[x]$ such that $f(x) \equiv G_1(x) \cdot H_1(x) (\bmod\ p)$.*

Then for any integer $i > 1$, there exist polynomials $G_i(x)$ and $H_i(x)$ in $\mathbb{I}_q[x]$, where $q = p^i$, such that $f(x) \equiv G_i(x) \cdot H_i(x) \bmod q$ with

$$G_i \equiv G_1 \bmod p$$
$$H_i \equiv H_1 \bmod p.$$

This lemma provides a method for lifting factors of a polynomial over a finite field $\mathbb{I}_p$ step-by-step towards a larger subdomain of $\mathbb{I}[x]$, while preserving the relevant properties, and eventually leads towards the desired result in $\mathbb{I}[x]$.

A variation of Hensel's lemma was proposed by Zassenhaus [1969].

5.5.2 **Zassenhaus–Hensel Lemma.** *Let p be a prime in $\mathbb{I}$ and $f(x)$ be a given polynomial in $\mathbb{I}[x]$. Let $G_1(x)$ and $H_1(x)$ be two relatively prime polynomials in $\mathbb{I}_p[x]$ such that*

$$f(x) \equiv G_1(x) \cdot H_1(x) (\bmod\ p).$$

Then for any integer $i > 1$, there exist polynomials $G_i(x)$ and $H_i(x)$ in $\mathbb{I}_q[x]$, where $q = p^{2^{i-1}}$, such that

$$f(x) \equiv G_i(x) \cdot H_i(x) \bmod q$$

with

$$G_i \equiv G_1 \bmod p$$
$$H_i \equiv H_1 \bmod p.$$

Note that the modulus increases here quadratically as i increases. Both Hensel's lemma and its variant are approximate methods in power series fields.

5.5.3 Remark. In fact Zassenhaus–Hensel lemma holds for any $q = p^{n^{i-1}}$. This leads to nth order convergent approximations.

5.5.4 *Relation to Newton's method.* Let $f(x) = c_0 + c_1 x + \cdots + c_n x^n$ be a polynomial whose coefficients are p-adic integers. Let $f'(x) = c_1 + \cdots + n c_n x^{n-1}$ be the derivative of $f(x)$. Let a_0 be a p-adic integer such that $f(a_0) \equiv 0 \bmod p$ and $f'(a_0) \not\equiv 0 \bmod p$. Then Hensel's lemma asserts that there exists a unique p-adic integer a such that

$$f(a) = 0 \quad \text{and} \quad a \equiv a_0 \bmod p.$$

Hensel's lemma is often called the p-adic Newton's lemma (Koblitz [1977]) as it provides an approximation technique that essentially mimics the Newton's method for finding a real root of a polynomial with real coefficients (Loos [1983], Lauer [1983]).

5.5.5 EXAMPLE. To solve $x^2 - 59x + 868 = 0$ in $\mathbb{Q}_5$ we first solve $(x^2 - 59x + 868) \bmod 5 = 0$ or $x^2 - 4x + 3 = 0$ for the first p-adic digit, namely $a_0 = b_0$ and obtain the higher order digits b_i by using the formula

$$b_i p^i = \frac{-f(a_{i-1})}{f'(a_{i-1})} \bmod p^{i+1}$$

where $a_i = \sum_{j=0}^{i} b_j p^j$.
    Thus we have

$$a_0 = 1$$

$$b_1 \cdot 5 = -\frac{1 - 59 + 868}{2 - 59} \bmod 25$$

$$= \tfrac{10}{7} \bmod 25 = 5.$$

Thus $b_1 = 1$, $a_1 = 6$;

$$b_2 \cdot 5^2 = -\frac{36 - 354 + 868}{12 - 59} = \frac{550}{47} \bmod 125$$

$$= 50 \cdot 8 \bmod 125 = 25.$$

Thus $b_2 = 1$ and $a_2 = 1 + 1 \cdot 5 + 1 \cdot 5^2 = 31_{ten}$ is a root of $f(x)$. Note that the correction term $[-f(a_{i-1})/f'(a_{i-1})] \bmod p^{i+1}$ mimics the correction term $[-f(x_{i-1})/f'(x_{i-1})]$ in Newton's method (5.1).

The p-adic Newton method is always guaranteed to converge to a root of a polynomial in $\mathbb{Q}_p$ (the function is always defined and the procedure terminates) unlike the real Newton's method (which may diverge, or oscillate and become undefined).

Note that if we had started with $a_0 = 3$, we would have obtained the other root $x = 28$ in $\mathbb{Q}_5$.

5.5.6 *Inversion of polynomial Hensel Code.* Let us for convenience write

$$H(p, r, \alpha(x)) = a(x) = \sum_{i=0}^{r-1} a_i x^i$$

$$H(p, r, \alpha(x)^{-1}) = c(x) = \sum_{i=0}^{r-1} c_i x^i,$$

where $c(x)$ is the multiplicative inverse of $a(x)$ over $\mathbb{I}_p[[x]]$. The NHZ method first constructs the starting approximation

(5.3) $b_{2^i}(x)(i = 0) = b_{2^0}(x) = a_0^{-1} \bmod p = c_0, \qquad a_0 \neq 0$

and performs the iteration

(5.4) $b_{2^i}(x) = b_{2^{i-1}}(x)[2 - a(x) \cdot b_{2^{i-1}}(x)] \bmod x^{2^i}$

for $i \geq 1$ until x^{2^i} is large enough to obtain r digit approximation $H(p, r, \alpha(x)^{-1})$.

The following theorem ensures the validity of this method.

5.5.7 **Theorem.** *There exists a polynomial sequence $\{b_{2^i}(x)\}$, $i \geq 0$ in $\mathbb{I}_p[[x]]$ such that*

$$a(x) \cdot b_{2^i}(x) = 1 \bmod x^{2^i} \quad \text{for all } i \geq 0,$$

where $a(x)$ is the polynomial whose inverse transform is to be obtained; or $b_{2^i}(x)$ is the inverse of $a(x) \bmod x^{2^i}$.

PROOF. We prove that the sequence $\{b_{2^i}\}, i \geq 0$ can be generated recursively and then prove by induction that it has the property stated, namely,

$$a(x) \cdot b_{2^i}(x) = 1 \bmod x^{2^i}.$$

The first member of the sequence (5.4), namely b_1, is obtained by using (5.3):

$$b_{2^0} = b_1 = a_0^{-1}.$$

Thus $a(x) \cdot a_0^{-1} \equiv 1 \bmod x^{2^0}$.

To prove that the theorem is true we use the fact that

$$a(x) \cdot b_{2^i}(x) \equiv 1 \bmod x^{2^i}$$

for $i = k-1$ $(k \geqslant 1)$; then by (5.4) we obtain

$$a(x) \cdot b_{2^k}(x) \equiv a(x) \cdot b_{2^{k-1}}(x)[2 - a(x) \cdot b_{2^{k-1}}(x)] \bmod x^{2^k}.$$

Since

$$a(x) \cdot b_{2^{k-1}}(x) \equiv 1 \bmod x^{2^{k-1}}$$
$$a(x) \cdot b_{2^{k-1}}(x) = 1 + x^{2^{k-1}} E_{k-1}(x),$$

where $E_{k-1}(x)$ is a residual polynomial. Thus we can write

$$a(x) \cdot b_{2^k}(x) \equiv (1 + x^{2^{k-1}} E_{k-1}(x))(1 - x^{2^{k-1}} E_{k-1}(x)) \bmod x^{2^k}$$
$$\equiv 1 \bmod x^{2^k}.$$

Since by construction the theorem holds for $k = 0$, it is true for all $k > 0$ by induction. □

Note that the proof above is based on induction; this is in contrast with the Newton's method for reals where the proof of convergence to a fixed point is based on contraction mapping; also there are no constraints for convergence.

5.5.8 EXAMPLE. Let $H(3, 2, (1+x)) = a(x) = (1, 1; 0)$.
To find $a(x)^{-1} \equiv c(x)$ we proceed as follows:

$$b_1(x) \equiv a_0^{-1} \bmod 3 = 1;$$
$$b_2(x) \equiv b_1(x) \cdot (2 - a(x)b_1(x)) \bmod x^2$$
$$= 1(2 - (1+x)) = 1 - x = 1 + 2x;$$
$$b_4(x) \equiv (1 + 2x)[2 - (1+x)(1+2x)] \bmod x^4$$
$$\equiv (1 + 2x)(1 + x^2) \bmod x^4$$
$$= (1 + 2x + x^2 + 2x^3);$$
$$b_8(x) \equiv (1 + 2x + x^2 + 2x^3)[2 - (1+x)(1 + 2x + x^2 + 2x^3)] \bmod x^8$$
$$\equiv (1 + 2x + x^2 + 2x^3)(1 + x^4) \bmod x^8$$
$$= 1 + 2x + x^2 + 2x^3 + x^4 + 2x^5 + x^6 + 2x^7.$$

Thus

$$H(3, 8, (1+x)^{-1}) = (1, 2, 1, 2, 1, 2, 1, 2; 0).$$

5.5.9 *Higher order convergence.* In Section 5.5.3 we remarked that Hensel's lemma can be extended for higher order convergence. We briefly mention here the extension of (5.4) to achieve higher order convergence rather than the quadratic convergence.

For this purpose we write (5.4) as

$$b_{2^i} = b_{2^{i-1}}[1 + [1 - ab_{2^{i-1}}]] \bmod x^{2^i}$$
$$= b_{2^{i-1}}[1 + d_{i-1}] \bmod x^{2^i},$$

where

(5.5) $$d_{i-1} = [1 - ab_{2^{i-1}}].$$

To obtain convergence of order $n > 2$, we replace the 2 in (5.5) by n so that $d_{i-1} = [1 - ab_{n^{i-1}}]$. Then we iterate, using the equation

(5.6) $$b_{n^i} = [b_{n^{i-1}}[1 + d_{i-1}(1 + d_{i-1}(1 + \cdots) \cdots)]] \bmod x^{n^i},$$

where there are $(n-1)$ terms in the nested expression.

For example, cubic convergence is obtained, if we use

(5.7) $$b_{3^i} = [b_{3^{i-1}}[1 + d_{i-1}(1 + d_{i-1})]] \bmod x^{3^i}.$$

5.6 Euclidean algorithm for finding $H(p, r, \alpha(x)^{-1})$

We already mentioned in Chapter I (Section 3.6.4) that the Euclidean gcd algorithm can be used to find the multiplicative inverse with respect to x^r of any polynomial $a(x)$ in $\mathbb{I}_p[x]$ provided $\gcd(a(x), x^r) = 1$. Since $H(p, r, \alpha(x)) \in \mathbb{I}_p[x]$, this algorithm can be directly used to compute $H(p, r, \alpha(x)^{-1})$ by using the seed matrix

$$\begin{bmatrix} x^r & 0 \\ H(p, r, \alpha(x)) & 1 \end{bmatrix}.$$

As already indicated, the remainder obtained in the last but one step in column 1 need not necessarily be the identity element 1. It could be any element c in $\mathbb{I}_p$. Accordingly, the result obtained should be multiplied by c^{-1} to obtain $\alpha(x)^{-1}$.

5.6.1 EXAMPLES.   (i)   Let $H(3, 2, (1 + x)) = a(x) = (1, 1; 0)$.   Find $a(x)^{-1} \bmod x^3$. We start with the seed matrix

$$\begin{bmatrix} x^3 & 0 \\ (1 + x) & 1 \end{bmatrix}$$

and obtain the following table:

	x^3	0
	$1 + x$	1
$x^2 - x + 1$	2	$2x^2 + x + 2$
$2x + 2$	0	$-x^3$

Thus

$$H(3, 3, (1 + x)^{-1}) = 2^{-1}(2x^2 + x + 2)$$
$$= x^2 + 2x + 1$$
$$= (1, 2, 1; 0).$$

(ii) Find $H(601, 5, (1+6x+6x^2)^{-1})$. Using computation similar to (i) above we obtain

$$H(601, 5, (1+6x+6x^2)^{-1}) = (1, 595, 30, 457, 83; 0).$$

(iii) Find $H(601, 5, (3+x+x^2)^{-1})$. Using computations similar to (i) we obtain

$$H(601, 5, (3+x+x^2)^{-1}) = (401, 267, 178, 52, 324; 0).$$

6 Forward and Inverse Mapping Algorithms

6.1 Forward mapping

The forward mapping from a rational polynomial $a(x)/b(x)$ where $a(x), b(x) \in \mathbb{I}_p[x]$ to $[a(x)b(x)^{-1}] \bmod x^r$ can be achieved either by computing the inverse transform of $b(x)$ and multiplying by $a(x)$ or directly by the single step method described in Section 3.6.4 of Chapter I.

Here we choose the seed matrix

$$\begin{bmatrix} x^r & 0 \\ b(x) & a(x) \end{bmatrix}$$

and directly compute $[a(x)b(x)^{-1}] \bmod x^r$.

6.1.1 EXAMPLE. Find $[(2+x)(1+x)^{-1}] \bmod x^3$ in $\mathbb{I}_3[[x]]$. The computations are shown below:

	x^3	0
	$(1+x)$	$(2+x)$
x^2-x+1	2	$2x^3+2x^2+x+1$
$2x+2$	0	$-x^4-2x^3$

Thus

$$(2+x)(1+x)^{-1} = 2^{-1}(1+x+2x^2) \bmod x^3$$
$$= 2+2x+x^2.$$

6.1.2 **Remark.** It is also possible to use the iterative inversion described in Section 5.5 to find $b(x)^{-1}$ and then compute $[a(x)b(x)^{-1}] \bmod x^r$.

6.2 Inverse mapping—Padé method

The polynomial Hensel Code is converted into a rational polynomial by the solution of a system of linear equations involving a matrix known as the Padé matrix.

Let us assume that

(6.1) $$H(p, 2R-1, \alpha(x)) = \sum_{i=0}^{2R-2} a_i x^i = a(x) \in F_p(x),$$

where $a_0 \neq 0$.

Let the rational polynomial

(6.2) $$\frac{c(x)}{d(x)} = \frac{c_0 + c_1 x + \cdots + c_{R-1} x^{R-1}}{d_0 + d_1 x + \cdots + d_{R-1} x^{R-1}}$$

uniquely correspond to $H(p, 2R-1, \alpha(x))$, where $c_i, d_i \in \mathbb{I}_p$ and $d_0 \neq 0$. We can conveniently rewrite (6.2) as

(6.3) $$\frac{c(x)}{d(x)} = \frac{d_0^{-1}(c_0 + c_1 x + \cdots c_{R-1} x^{R-1})}{1 + d_0^{-1}(d_1 x + \cdots + d_{R-1} x^{R-1})},$$

where each $(c_i \cdot d_0^{-1})$ and $(d_i \cdot d_0^{-1}) \in \mathbb{I}_p$.

Let us rewrite (6.3) as

(6.4) $$\frac{c(x)}{d(x)} = \frac{c_0^1 + c_1^1 x + \cdots + c_{R-1}^1 x^{R-1}}{1 + d_1^1 x + \cdots + d_{R-1}^1 x^{R-1}}.$$

Our aim is to determine $c_i^1 (i = 0, 1, 2, \ldots, R-1)$ and $d_i^1 (i = 1, 2, \ldots, (R-1))$. Using (6.4) and (6.1) we have

(6.5) $c_0^1 + c_1^1 x + \cdots + c_{R-1}^1 x^{R-1}$

$$= (1 + d_1^1 x + \cdots + d_{R-1}^1 x^{R-1}) \cdot (a_0 + a_1 x + \cdots + a_{2R-2} x^{2R-2}).$$

Equating the coefficients of like powers in (6.5) we obtain a system of linear equations of the form

(6.6) $$Ad = c,$$

where $c^T = [c_0^1, c_1^1, \ldots, c_{R-1}^1, 0, \ldots, 0]$ and $d^T = [1, d_1^1, \ldots, d_{R-1}^1, 0, \ldots, 0]$ are both $(2R-1) \times 1$ vectors and A is a lower triangular matrix of order $(2R-1) \times (2R-1)$ given by

$$A = \begin{bmatrix}
a_0 & 0 & 0 & 0 & \cdots & 0 \\
a_1 & a_0 & 0 & 0 & \cdots & 0 \\
a_2 & a_1 & a_0 & 0 & \cdots & 0 \\
\cdots & \cdots & \cdots & \cdots & \cdots & \\
a_{R-1} & a_{R-2} & \cdots & \cdots & \cdots & a_0 \\
a_R & a_{R-1} & \cdots & \cdots & \cdots & a_0 \\
\cdots & \cdots & \cdots & \cdots & \cdots & \\
a_{2R-2} & a_{2R-3} & \cdots & \cdots & \cdots & a_0
\end{bmatrix}$$

We call the matrix A "Padé" matrix in honour of Padé who suggested a rational approximation scheme for a formal power series (Baker [1975], Baker and Grave Morris [1981]). The Padé matrix relates the formal power series to a rational polynomial by (6.6). This matrix has the following properties:

(i) A is a circulant or a Toeplitz matrix with the property $a_{ij} = a_{i+1,j+1}$; $i, j = 1, 2, \ldots, (n-1)$ (see Davis [1979]).

(ii)   The elements of the first column are the coefficients of Hensel Code in increasing degree terms and the successive columns are shifted Hensel Codes (by one row down).

(iii)   The elements of the last row are the coefficients of the Hensel Code in decreasing degree terms and the successive rows from bottom up are left-shifted (by one column) Hensel Codes.

(iv)   The last $(R-1)$ rows enable us to solve for d_i^1 $(i = 1, 2, \ldots, R-1)$. (If the Hensel Code is not unique, we cannot obtain a unique solution.)

(v)   The d_i^1's are then fed into (6.6) and a matrix multiplication Ad^1 involving the first R rows yields c_i^1 $(i = 0, 1, \ldots, R-1)$.

The vectors c^T and d^T thus obtained will be over $\mathbb{I}_p$. These are to be converted to rationals in the usual manner by the inverse mapping algorithm based on Euclidean gcd algorithm (Section 4.8.2). The factor d_0^{-1} occurring as a common factor in (6.3) can then be cancelled and the rational polynomial brought to the order-N, degree $(R-1/R-1)$ Padé form

$$\frac{c(x)}{d(x)} = \frac{c_0 + c_1 x + \cdots + c_{R-1} x^{R-1}}{d_0 + d_1 x + \cdots + d_{R-1} x^{R-1}},$$

where $0 \leqslant |c_i|, |d_i| \leqslant N$.

6.2.1 **Remark.** For an $O(n \log^2 n)$ in time and $O(n)$ in space (where $n = R - 1$), fast algorithm for the solution of Toeplitz systems and computation of n/n Padé approximants, see Brent, Gustavson, Yun [1980]. This is a very useful paper which relates the solution of Toeplitz systems to the Euclidean algorithm.

6.2.2 EXAMPLES. (i) Convert $H(601, 5, \alpha(x)) = (1, 600, 2, 595, 24; 0)$ assuming the form (6.4):

$$
\begin{bmatrix}
1 & 0 & 0 & 0 & 0 \\
600 & 1 & 0 & 0 & 0 \\
2 & 600 & 1 & 0 & 0 \\
595 & 2 & 600 & 1 & 0 \\
24 & 595 & 2 & 600 & 1
\end{bmatrix}
\begin{bmatrix}
1 \\
d_1^1 \\
d_2^1 \\
0 \\
0
\end{bmatrix}
=
\begin{bmatrix}
c_0^1 \\
c_1^1 \\
c_2^1 \\
0 \\
0
\end{bmatrix}
$$

we solve

$$2d_1^1 + 600d_2^1 = -595 = 6$$
$$595d_1^1 + 2d_2^1 = -24$$

and obtain $d_1^1 = 6$; $d_2^1 = 6$. Using these we get $c_0^1 = 1$, $c_1^1 = 5$, and $c_2^1 = 2$. Thus the rational is

$$\alpha(x) = \frac{1 + 5x + 2x^2}{1 + 6x + 6x^2} \in P(2/2, 10, x).$$

In this case, the coefficients $\in \mathbb{I}_{601}$, after conversion to P_N using Algorithm 4.8.2, result in integers.

(ii) Convert $H(601, 5, \alpha(x)) = (403, 267, 179, 252, 257; 0)$ assuming the form (6.4):

$$
\begin{bmatrix}
403 & 0 & 0 & 0 & 0 \\
267 & 403 & 0 & 0 & 0 \\
179 & 267 & 403 & 0 & 0 \\
252 & 179 & 267 & 403 & 0 \\
257 & 252 & 179 & 267 & 403
\end{bmatrix}
\begin{bmatrix}
1 \\ d_1^1 \\ d_2^1 \\ 0 \\ 0
\end{bmatrix}
=
\begin{bmatrix}
c_0^1 \\ c_1^1 \\ c_2^1 \\ 0 \\ 0
\end{bmatrix}
$$

we solve

$$179d_1^1 + 267d_2^1 = -252$$
$$252d_1^1 + 179d_2^1 = -257$$

and obtain

$$d_0^1 = 1, \qquad d_1^1 = 401, \qquad d_2^1 = 401$$
$$c_0^1 = 403, \qquad c_1^1 = 201, \qquad c_2^1 = 202$$

which when converted to P_N using Algorithm 4.8.2 gives

$$\alpha(x) = \frac{\frac{7}{3} + \frac{2}{3}x + \frac{5}{3}x^2}{1 + \frac{1}{3}x + \frac{1}{3}x^2}.$$

On cancellation of the common denominator we obtain

$$\frac{7 + 2x + 5x^2}{3 + x + x^2} \in P(2/2, 10, x).$$

(iii) Convert $H(601, 5, \alpha(x)) = (1, 595, 30, 457, 83; 0)$ assuming the form (6.4):

$$
\begin{bmatrix}
1 & 0 & 0 & 0 & 0 \\
595 & 1 & 0 & 0 & 0 \\
30 & 595 & 1 & 0 & 0 \\
457 & 30 & 595 & 1 & 0 \\
83 & 457 & 30 & 595 & 1
\end{bmatrix}
\begin{bmatrix}
1 \\ d_1^1 \\ d_2^1 \\ 0 \\ 0
\end{bmatrix}
=
\begin{bmatrix}
c_0^1 \\ c_1^1 \\ c_2^1 \\ 0 \\ 0
\end{bmatrix}
$$

we solve

$$30d_1^1 + 595d_2^1 = -457$$
$$457d_1^1 + 30d_2^1 = -83$$

and obtain

$$d_1^1 = 6, \qquad d_2^1 = 6,$$
$$c_0^1 = 1, \qquad c_1^1 = 0, \qquad c_2^1 = 0.$$

Thus

$$\alpha(x) = \frac{1}{1 + 6x + 6x^2} \in P(2/2, 10, x).$$

(iv) Convert $H(601, 5, \alpha(x)) = (5, 6, 6, 0, 0; 0)$ assuming the form (6.4):

$$\begin{bmatrix} 5 & 0 & 0 & 0 & 0 \\ 6 & 5 & 0 & 0 & 0 \\ 6 & 6 & 5 & 0 & 0 \\ 0 & 6 & 6 & 5 & 0 \\ 0 & 0 & 6 & 6 & 5 \end{bmatrix} \begin{bmatrix} 1 \\ d_1^1 \\ d_2^1 \\ 0 \\ 0 \end{bmatrix} = \begin{bmatrix} c_0^1 \\ c_1^1 \\ c_2^1 \\ 0 \\ 0 \end{bmatrix}$$

we solve

$$6d_1^1 + 6d_2^1 = 0$$
$$6d_2^1 = 0$$

and obtain $d_2^1 = d_1^1 = 0$; $c_0^1 = 5, c_1^1 = 6$; $c_2^1 = 6$. Thus $\alpha(x) = 5 + 6x + 6x^2 \in P(2/2, 10, x)$.

(v) Convert $H(601, 5, \alpha(x)) = (1, 595, 36, 385, 94; 0)$ assuming the form (6.4):

$$\begin{bmatrix} 1 & 0 & 0 & 0 & 0 \\ 595 & 1 & 0 & 0 & 0 \\ 36 & 595 & 1 & 0 & 0 \\ 385 & 36 & 595 & 1 & 0 \\ 94 & 385 & 36 & 595 & 1 \end{bmatrix} \begin{bmatrix} 1 \\ d_1^1 \\ d_2^1 \\ 0 \\ 0 \end{bmatrix} = \begin{bmatrix} c_0^1 \\ c_1^1 \\ c_2^1 \\ 0 \\ 0 \end{bmatrix}$$

we solve

$$36d_1^1 - 6d_2^1 = -385$$
$$385d_1^1 - 36d_2^1 = -94$$

which gives $d_1^1 = 6$, $d_2^1 = 0$ and $c_0^1 = 1$, $c_1^1 = 0$, $c_2^1 = 0$. Thus

$$\alpha(x) = \frac{1}{1 + 6x} \in P(2/2, 10, x).$$

6.3 Euclidean inverse mapping

The algorithm earlier described in Chapter I, Gregory and Krishnamurthy [1984] for the inverse mapping of a Hensel Code (of a rational number) to its equivalent Farey rational can be extended to the case of polynomial

Hensel Codes. This is not surprising since both integers and polynomials (over a field) belong to the gcd domain and the algorithm used for integers can be directly extended to polynomials, if the Euclidean degree function is defined suitably. However, the proof of such an extended algorithm requires the definition of a suitable norm (or valuation) for the polynomials and the application of this norm to extend all the theorems concerning the continued fraction of rational numbers to the continued fraction of rational functions over a finite field.

The main theorem we have to prove in this case is that the Extended Euclidean algorithm starting with the seed matrix (Section 3.6.4, Chapter I)

$$\begin{bmatrix} x^r & 0 \\ H(p, r, \alpha(x)) & 1 \end{bmatrix}$$

generates the Padé rational $P(R-1/R-1, F_p(x))$ represented by the given polynomial Hensel Code $H(p, r, \alpha(x))$, where $r = 2R - 1$. (Of course, this rational polynomial is then to be converted to $P(R-1/R-1, N, x)$ form using the Euclidean inverse mapping method for each one of the coefficients in the numerator and the denominator.) In other words, we need to prove the analogues of Theorems 6.38 and 6.39 of Chapter I of Gregory and Krishnamurthy [1984] for the case of rational polynomial functions over a field. For this purpose we now recall some of the important properties (theorems) of continued fractions from Hardy and Wright [1979] and show that these can be extended to polynomials belonging to $\mathbb{I}_p[x]$, by using the concept of a norm or a valuation.

6.3.1 *Finite continued fractions over* $\mathbb{I}_p[x]$. We shall describe the function

$$A_0(x) + \cfrac{1}{A_1(x) + \cfrac{1}{A_2(x) + \cfrac{1}{A_3(x) + \cdots \cfrac{}{+ \cfrac{1}{A_N(x)}}}}}$$

as a finite continued fraction and denote this by

$$[A_0(x), A_1(x), \ldots, A_N(x)],$$

where $A_i(x) \in \mathbb{I}_p[x]$ $(i = 0, 1, \ldots, N)$ are partial quotients; for convenience, we further omit the indeterminate x and write $A_i(x)$ as simply A_i.

If P_n and Q_n are defined by

$$P_0 = A_0, \qquad P_1 = A_1 A_0 + 1, \ldots, P_n = A_n P_{n-1} + P_{n-2}, \qquad (2 \leq n \leq N)$$

and

$$Q_0 = 1, \qquad Q_1 = A_1, \ldots, Q_n = A_n Q_{n-1} + Q_{n-2}, \qquad (2 \le n \le N)$$

then

$$[A_0, A_1, \ldots, A_n] = P_n / Q_n$$

is the nth convergent to

$$[A_0, A_1, \ldots, A_N].$$

The nth convergent satisfies the following properties:

(6.7) (i) $P_n Q_{n-1} - Q_n P_{n-1} = (-1)^{n-1}$

(6.8) (ii) P_n / Q_n is in its lowest terms but for a constant of proportionality $\in \mathbb{I}_p$.

6.3.2 *Relation to Euclidean algorithm.* Let us now consider the analogue of the seed matrix (6.34) of Chapter I, Gregory and Krishnamurthy [1984], of the Euclidean algorithm:

$$(6.9) \qquad \begin{bmatrix} a_{-1} & b_{-1} & c_{-1} \\ a_0 & b_0 & c_0 \end{bmatrix} = \begin{bmatrix} x^r & 0 & -1 \\ H(p, r, \alpha(x)) & 1 & 0 \end{bmatrix}$$

and let us define the sequence of triples $\{a_i, b_i, c_i\}$ by the recursion, while $a_{i-1} \ne 0$:

$$(6.10) \qquad q_i = \text{quotient} \left[\frac{a_{i-2}}{a_{i-1}} \right]$$

$$(6.11) \qquad a_i = a_{i-2} - q_i a_{i-1}$$

$$(6.12) \qquad b_i = b_{i-2} - q_i b_{i-1}$$

$$(6.13) \qquad c_i = c_{i-2} - q_i c_{i-1}$$

for $i = 1, 2, \ldots, (N+1)$.

It is easily seen that the sequence

$$\left\{ \frac{P_0}{Q_0} = \frac{b_1}{c_1}, \frac{P_1}{Q_1} = \frac{b_2}{c_2}, \ldots, \frac{P_N}{Q_N} = \frac{b_{N+1}}{c_{N+1}} \right\}$$

is the complete sequence of continued fraction convergents of $x^r / H(p, r, \alpha(x))$.

Also the following properties hold:

(6.14) $a_i = H(p, r, \alpha(x)) \cdot b_i - x^r c_i \qquad (i = 1, 2, \ldots, N+1)$

(6.15) $\deg c_i + \deg a_{i-1} = 2R - 2, \qquad 1 \le i \le N+1$

(6.16) $\deg b_i + \deg a_{i-1} = 2R - 1, \qquad 0 \le i \le N+1$

(6.17) $\deg c_i > \deg c_{i-1}, \qquad i \ge 2$

(6.18) $\deg b_i > \deg b_{i-1}, \qquad i \ge 1$

(6.19) $\deg Q_i > \deg Q_{i-1}, \qquad i \ge 1$

(6.20) $\deg P_i > \deg P_{i-1}, \qquad i \ge 0.$

The following example illustrates the above properties.

6.3.3 EXAMPLE. Let

$$\begin{bmatrix} a_{-1} & b_{-1} & c_{-1} \\ a_0 & b_0 & c_0 \end{bmatrix} = \begin{bmatrix} x^3 & 0 & -1 \\ 1+16x+2x^2 & 1 & 0 \end{bmatrix},$$

where $a_i, b_i, c_i \in \mathbb{I}_{17}[x]$.
  Then we get the sequence

i	q_i	a_i	b_i	c_i
-1	—	x^3	0	-1
0	—	$1+16x+2x^2$	1	0
1	$9x+13$	$4x+4$	$8x+4$	-1
2	$9x+12$	4	$13x^2+4x+4$	$9x+12$
3	$x+1$	0	$4x^3$	$8x^2+13x+4$

It is easy to see that

$$\frac{ux^3}{1+16x+2x^2} = (9x+13) + \cfrac{1}{(9x+12) + \cfrac{1}{(x+1)}},$$

where $u \neq 0$ is a unit in $\mathbb{I}_p[x]$; here $u = 13 \in \mathbb{I}_{17}$.
  The successive convergents are:

$$\left\{ \frac{8x+4}{-1} = 9x+13, \quad \frac{13x^2+4x+4}{9x+12}, \quad \frac{4x^3}{8x^2+13x+4} = \frac{x^3}{2x^2+16x+1} \right\}.$$

Also note

$$P_n Q_{n-1} - P_{n-1} Q_n = (-1)^{n-1}.$$

6.3.4 *Valuation or norm.* We now introduce the concept of a valuation or norm for the rational polynomial functions: For all $a(x)/b(x) \in F_p(x)$

$$(6.21) \qquad \phi(a(x)/b(x)) \begin{cases} = {}_2 \deg a(x) - \deg b(x), & \text{for } a(x) \neq 0 \\ = 0, & \text{otherwise,} \end{cases}$$

where deg denotes the degree function (Section 3.2, Chapter I) satisfying the following properties:

$$(6.22) \qquad \begin{aligned} \deg[a(x) \cdot b(x)] &= \deg a(x) + \deg b(x) \\ \deg[a(x) + b(x)] &\leq \max\{\deg a(x), \deg b(x)\}. \end{aligned}$$

It is easy to see that $\phi(\cdot)$ is a norm or valuation, (Bachmann [1964],

Jacobson [1976]), since

(i) $\phi(a(x)/b(x)) = 0, \quad \text{iff } a(x)/b(x) = 0$

(ii) $\phi\left[\dfrac{a(x)}{b(x)} \cdot \dfrac{c(x)}{d(x)}\right] = \phi\left[\dfrac{a(x)}{b(x)}\right] \cdot \phi\left[\dfrac{c(x)}{d(x)}\right]$

(iii) $\phi\left[\dfrac{a(x)}{b(x)} + \dfrac{c(x)}{d(x)}\right] \leqslant \max\left\{\phi\left[\dfrac{a(x)}{b(x)}\right], \phi\left[\dfrac{c(x)}{d(x)}\right]\right\}.$

We now prove the following theorem (analogue of Theorem 172, p. 140, Hardy and Wright [1979].

6.3.5 **Theorem.** *If* $\alpha = (P\beta + R)/(Q\beta + S)$ *where* $\phi(\beta) > 1$ *and* P, Q, R, S *are polynomials* $\in \mathbb{I}_p[x]$ *such that*

$$PS - QR = \pm 1, \qquad \phi(Q) > \phi(S) > 0,$$

then R/S *and* P/Q *are two consecutive convergents to the simple continued fraction whose value is* α. *If* R/S *is the* $(n-1)$th *convergent and* P/Q *the* nth, *then* β *is the* $(n+1)$th *complete quotient.*

PROOF. Develop P/Q as a simple continued fraction

(6.23) $P/Q = [A_0, A_1, \ldots, A_n] = P_n/Q_n.$

We may suppose that n is odd or even since it is always possible to write

$$[A_0, A_1, \ldots, A_n] = [A_0, A_1 \ldots, A_n - c, c^{-1}]$$

for some $c(\neq 0) \in \mathbb{I}_p$.

Thus we can choose n such that

(6.24) $PS - QR = \pm 1 = (-1)^{n-1}.$

Note that $\gcd(P, Q) = 1$ and P_n and Q_n satisfying (6.7) and (6.8) namely:

$$P_n Q_{n-1} - Q_n P_{n-1} = (-1)^{n-1}$$

and P_n/Q_n is in its lowest terms (but for a constant of proportionality $\in \mathbb{I}_p$).

From (6.23) we have $P = P_n$ and $Q = Q_n$ so that using (6.24) we get

$$P_n S - Q_n R = PS - QR = (-1)^{n-1}$$

$$= P_n Q_{n-1} - Q_n P_{n-1}$$

or $P_n(S - Q_{n-1}) = Q_n(R - P_{n-1})$. Since $\gcd(P_n, Q_n) = 1$, Q_n divides $(S - Q_{n-1})$. But we are given by the condition of theorem that

$$\phi(Q_n) = \phi(Q) > \phi(S);$$

this implies $\deg S < \deg Q_n$.

Also by (6.19) we have

$$\deg Q_n > \deg Q_{n-1}.$$

Hence, if Q_n divides $(S - Q_{n-1})$ then necessarily $S - Q_{n-1} = 0$; or $S = Q_{n-1}$ and $R = P_{n-1}$. Thus

$$\alpha = \frac{P_n \beta + P_{n-1}}{Q_n \beta + Q_{n-1}}$$

that is

$$\alpha = [A_0, A_1, \ldots, A_n, \beta].$$

If now we develop β as a continued fraction, we obtain

$$\beta = [A_{n+1}, A_{n+2}, \cdots],$$

where $\phi(A_{n+1}) \geqslant 1$. Hence

$$\alpha = [A_0, A_1, \ldots, A_n, A_{n+1}, \cdots]$$

is a continued fraction with $P_{n-1}/Q_{n-1} = R/S$ and $P_n/Q_n = P/Q$ respectively, as the consecutive convergents of α, and β the $(n+1)$th complete quotient.   □

We now prove the analogue of Theorem 6.38 Chapter I, Gregory and Krishnamurthy [1984] for rational functions over $F_p(x)$.

6.3.6 **Theorem.** *If* $\phi(P/Q - \alpha) \leqslant [1/2\phi(Q^2)]$ *then* P/Q *is a continued fraction convergent over* α.

PROOF. Set

(6.25) $\theta = Q^2(P/Q - \alpha).$

We can express P/Q as a finite continued fraction, namely $P/Q = [A_0, A_1, \ldots, A_n]$ and since we can make n odd or even at our discretion (see Hardy and Wright [1979], p. 133, Theorem 158) we may suppose

(6.26) $P_n Q_{n-1} - P_{n-1} Q_n = (-1)^{n-1} = \Delta$ (say).

We can then write

(6.27) $\alpha = \frac{w P_n + P_{n-1}}{w Q_n + Q_{n-1}},$

where P_n/Q_n and P_{n-1}/Q_{n-1} are the last and last but one convergents to the continued fraction for P/Q. Then from (6.25)

$$\theta/Q_n^2 = (P_n/Q_n - \alpha)$$

(6.28) $= \frac{P_n Q_{n-1} - P_{n-1} Q_n}{Q_n(w Q_n + Q_{n-1})}$

$$= \frac{\Delta}{Q_n(w Q_n + Q_{n-1})}.$$

Hence it follows that

(6.29)
$$\frac{(wQ_n + Q_{n-1})}{Q_n} = \frac{\Delta}{\theta}.$$

By (6.19)

$$\deg Q_n > \deg Q_{n-1}$$

or

$$\phi(Q_n) > \phi(Q_{n-1})$$

or

(6.30)
$$\phi\left[\frac{Q_{n-1}}{Q_n}\right] < 1.$$

But from (6.29)

(6.31)
$$\phi\left[\frac{1}{\theta}\right] = \phi\left[\frac{wQ_n + Q_{n-1}}{Q_n}\right]$$
$$= \phi\left[w + \frac{Q_{n-1}}{Q_n}\right]$$
$$\leq \max[\phi(w), \phi(Q_{n-1}/Q_n)]$$
$$\leq \max[\phi(w), x],$$

where $x \leq 1$ by (6.30). But by hypothesis, $\phi(\theta) \leq 1/2$ or $\phi[1/\theta] \geq 2$; this implies

(6.32)
$$\phi(w) \geq 2.$$

By application of Theorem 6.3.5 we see immediately P_{n-1}/Q_{n-1} and P_n/Q_n are two consecutive convergents to α and $P/Q = P_n/Q_n$.   □

We are now ready to prove the analogue of the main theorem 6.39 of Chapter I, Gregory and Krishnamurthy [1984], which permits the conversion of polynomial Hensel Code.

6.3.7 **Theorem** (Conversion Algorithm). *If $a(x)/b(x) \in F_p(x)$ such that $\deg(a(x), b(x)) \leq (R-1)$, $\gcd(a(x), b(x)) = 1$, and $\gcd(b(x), x) = 1$ and $r \geq (2R-1)$ and $a(x)b^{-1}(x) \bmod x^r = H(p, r, \alpha(x)) = k(x)$, then there is an i such that $(a(x), b(x)) = (a_i, b_i)$, where $\{(a_i, b_i)\}$ is generated by (6.10) to (6.13) starting with the seeding matrix (6.9).*

PROOF. We only need to prove that, if $r \geq 2R - 1$, the sufficiency condition in Theorem 6.3.6 holds. By the statement of the theorem, we have

(6.33)
$$a(x) \equiv b(x)k(x) - x^r t(x)$$

for some $t(x) \in \mathbb{I}_p[x]$. Thus

$$\phi\left[\frac{k(x)}{x^r} - \frac{t(x)}{b(x)}\right] = \phi\left[\frac{a(x)}{x^r b(x)}\right]$$

$$= \frac{\phi(a(x))}{\phi(x^r)\phi(b(x))}$$

$$= \frac{\phi(a(x))\phi(b(x))}{\phi(x^r)(\phi(b(x)^2))}$$

$$\leqslant 2^{2R-2}/[2^{2R-1}\phi(b(x)^2)]$$

$$\leqslant \frac{1}{2\phi(b(x)^2)}.$$

Therefore using Theorem 6.3.6, we deduce that $t(x)/b(x)$ is a continued fraction convergent of $k(x)/x^r$.

Since the sequence

$$\left\{\frac{P_0}{Q_0} = \frac{b_1}{c_1}, \frac{P_1}{Q_1} = \frac{b_2}{c_2}, \ldots, \frac{P_N}{Q_N} = \frac{b_{N+1}}{c_{N+1}}\right\}$$

is the complete sequence of continued fraction convergents of $x^r/H(p, r, \alpha(x)) = x^r/k(x)$, it follows that

$$\left\{\frac{0}{1}, \frac{c_1}{b_1}, \frac{c_2}{b_2}, \ldots, \frac{c_{N+1}}{b_{N+1}}\right\}$$

is the sequence of convergents of $k(x)/x^r$. Hence there exists an i, where $1 \leqslant i \leqslant N+1$ such that $t(x)/b(x) = c_i/b_i$.

Finally from (6.14) we have,

$$a_i = k(x)b_i - x^r c_i$$

or

$$\frac{a_i}{b_i} = k(x) - x^r \frac{c_i}{b_i}$$

and from (6.33)

$$a(x) = k(x)b(x) - x^r t(x) \quad \text{or} \quad \frac{a(x)}{b(x)} = k(x) - x^r \frac{t(x)}{b(x)}.$$

Hence $a_i/b_i = a(x)/b(x)$ and so $(a(x), b(x)) = (a_i, b_i)$.   □

6.3.8 **Remark.** McEliece and Shearer [1978] give a different proof to show that the Padé approximant can be computed by Euclid's algorithm. Our proof essentially follows the argument used by Hardy and Wright [1979] for integers and generalizes the earlier proof given in Gregory and Krishnamurthy [1984] for the conversion of Hensel Codes to their equivalent Farey rationals.

6.3.9 EXAMPLES. (i) Convert $H(17, 3, \alpha(x)) = (1, 16, 2; 0)$ with $R = 2$. The Euclidean inverse mapping is given in the following table:

	x^3	0
	$1 + 16x + 2x^2$	1
$9x + 13$	$4x + 4$	$8x + 4$
$9x + 12$	4	$13x^2 + 4x + 4$
$x + 1$	0	$4x^3$

In the table above the rational polynomial which satisfies the condition that the degrees of the numerator and denominator are both $\leqslant R - 1 = 1$ is the polynomial $(4x + 4)/(8x + 4)$.

We convert this into the form

$$\frac{(4x + 4) \cdot 4^{-1}}{(8x + 4) \cdot 4^{-1}} = \frac{x + 1}{2x + 1} \in P(1/1, F_{17}(x)).$$

We then map the coefficients in numerator and denominator to Padé rationals using Algorithm 4.8.2. In this case, these turn out to be integers and we obtain

$$\alpha(x) = \frac{x + 1}{2x + 1} \in P(1/1, 2, x).$$

(ii) Convert $H(601, 5, \alpha(x)) = (1, 600, 2, 595, 24; 0)$. The computations are shown below. We stop at the second step as the degree conditions are met with $(R - 1 = 2)$.

	x^5	0
	$24x^4 + 595x^3 + 2x^2 + 600x + 1$	1
$-25x + 144$	$313x^3 - 313x^2 + 169x - 144$	$25x - 144$
$50x - 263$	$-16x^2 - 40x - 8$	$-48x^2 - 48x - 8$

Thus

$$\alpha(x) = \frac{-16x^2 - 40x - 8}{-48x^2 - 48 - 8}$$

$$= \frac{2x^2 + 5x + 1}{6x^2 + 6x + 1} \in P(2/2, F_{601}(x))$$

which when converted to Padé rationals through the inverse mapping Algorithm 4.8.2 for each coefficient leads to

$$\alpha(x) = \frac{2x^2 + 5x + 1}{6x^2 + 6x + 1} \in P(2/2, 10, x).$$

6.4 Common denominator method

The common denominator method described in Gregory and Krishnamurthy [1984], Chapter I, is easily extended for the conversion of polynomial Hensel Code. Suppose $\alpha(x) = a(x)/b(x) \in P(R-1/R-1, \hat{F}_p(x))$ (Padé rational polynomial) and $H(p, r, \alpha(x)) = [a(x) \cdot b(x)^{-1}]$ mod $x^r = k(x)$. We can map $k(x)$ onto $\alpha(x)$ as follows:

If a multiple $t(x)b(x)$ can be found with $t(x) \in P(R-1/R-1, \hat{F}_p(x))$ then $t(x)a(x) = [t(x)b(x)k(x)]$ mod x^r and thus

$$\alpha(x) = \frac{t(x)a(x)}{t(x)b(x)}.$$

Note, however, $\alpha(x)$ may not be in the lowest or reduced form. This method enables us to obtain a rational polynomial in which the numerator and the denominator belong to $I_p[x]$. The coefficients are then converted to the rationals using Algorithm 4.8.2.

6.4.1 EXAMPLE. Consider $H(3, 3, \alpha(x)) = (2, 2, 1; 0)$ and a common denominator $t(x)b(x) = (1 + x^2 + 2x)$. We have $k(x) = (2 + 2x + x^2)$ and

$$t(x)a(x) = [(1 + x^2 + 2x) \cdot (2 + 2x + x^2)] \bmod x^3$$
$$= (2 + x^2).$$

Thus

$$\alpha(x) = \frac{(2 + x^2)}{(1 + x^2 + 2x)} \notin P(1/1, F_3(x)).$$

After reduction

$$\alpha(x) = (2 + x)/(1 + x) \in P(1/1, F_3(x)).$$

7 Direct Solution of Linear Systems and Matrix Inversion

The polynomial Hensel Codes will find applications in solving linear system of equations whose coefficient matrix has polynomial entries and inverting polynomial matrices. We will illustrate these applications with examples.

7.1 Gaussian elimination (with scaling)

Since all the basic arithmetic operations are defined for Hensel Codes it is possible to use Gaussian elimination (with scaling) for solving linear systems with rational polynomial coefficients, in a manner analogous to that described in Section 7.5, Chapter II, Gregory and Krishnamurthy

[1984]. For performing Gaussian elimination, however, it is necessary to have an estimate of the coefficient size and degree of the rational polynomial elements of the inverse matrix or the solution vector. We described earlier in Sections 2.6.8 and 2.6.9 of Chapter II, how to obtain these estimates. Let $(R - 1)$ be the maximum degree of the numerator and denominator polynomials and N be the coefficient size obtained by these estimates. Then we need to choose a prime p (or a power of a prime p^r) such that

$$p \geqslant 2RN^2 + 1$$

with the code length (maximal degree) of the Hensel Code being $(2R - 1)$.

7.1.1 EXAMPLE. Let us solve the system of linear algebraic equations $Ay = g$ where

$$y = \begin{bmatrix} y_1 \\ y_2 \\ y_3 \end{bmatrix},$$

$$A = \begin{bmatrix} x & x & (x-1) \\ (x^2+x) & (x^2+2x) & (x^2-1) \\ (2x^2-2x) & (x^2-2x) & (2x^2-3x+2) \end{bmatrix} \in \mathbb{Q}(x)$$

and

$$g = \begin{bmatrix} -3x^3 & + & 5x^2 & + & x \\ -3x^4 & + & 2x^3 & + & 8x^2 & + & x \\ -6x^4 & + & 11x^3 & - & 8x^2 & - & 2x \end{bmatrix} = \begin{bmatrix} g_1 \\ g_2 \\ g_3 \end{bmatrix}.$$

We begin by writing the augmented matrix (A,g) in Hensel Code. We choose $p = 101$ (as it is adequate for the present example) and $(2R - 1) = 9$; however, to conserve space and for clarity, we omit the trailing zeros, in the following example.

This gives us $[A,g] =$

$$\begin{bmatrix} (0, 1; 0) & (0, 1; 0) & (100, 1; 0) & (0, 1, 5, 98; 0) \\ (0, 1, 1; 0) & (0, 2, 1; 0) & (100, 0, 1; 0) & (0, 1, 8, 2, 98; 0) \\ (0, 99, 2; 0) & (0, 99, 1; 0) & (2, 98, 2; 0) & (0, 99, 93, 11, 95; 0) \end{bmatrix}.$$

Using Gaussian elimination (with scaling), we obtain the matrices

$$\begin{bmatrix} (1; 0) & (1; 0) & (100, 1; -1) & (1, 5, 98; 0) \\ (0; 0) & (0, 1; 0) & (0; 0) & (0, 0, 2; 0) \\ (0; 0) & (0, 0, 100; 0) & (0, 1; 0) & (0, 0, 0, 96; 0) \end{bmatrix}$$

$$\begin{bmatrix} (1; 0) & (1; 0) & (100, 1; -1) & (1, 5, 98; 0) \\ (0; 0) & (1; 0) & (0; 0) & (0, 2; 0) \\ (0; 0) & (0; 0) & (0, 1; 0) & (0, 0, 0, 98; 0) \end{bmatrix}$$

$$\begin{bmatrix} (1; 0) & (1; 0) & (100, 1; -1) & (1, 5, 98; 0) \\ (0; 0) & (1; 0) & (0; 0) & (0, 2; 0) \\ (0; 0) & (0; 0) & (1; 0) & (0, 0, 98; 0) \end{bmatrix}.$$

The product of the pivots gives

$$(0, 1; 0) \cdot (0, 1; 0) \cdot (0, 1; 0) = x^3$$

and so $\det A = x^3$.

To solve the triangular system of equations we use back substitution. Thus we obtain

$$y_3 = (0, 0, 98; 0) = -3x^2$$
$$y_2 = (0, 2; 0) \quad\;\; = 2x$$
$$y_1 = (1; 0) \qquad = 1.$$

7.2 Matrix inversion using reduction

We now use the reduction algorithm described in Section 4.5 of Chapter III, Gregory and Krishnamurthy [1984] for matrix inversion. Here we reduce the matrix to $\begin{bmatrix} I_r & 0 \\ 0 & 0 \end{bmatrix}$ form (r is the rank of the given matrix A) by using elementary column and row transformations (Chapter II).

First we reduce A to M_1 in simple row-echelon form. Obviously M_1^T is in simple column echelon form. We then transform M_1^T to the form $\begin{bmatrix} I_r & 0 \\ 0 & 0 \end{bmatrix} = U$, where r is the rank of A. If A is nonsingular, then $U = I$; thus,

$$EA = M_1$$
$$FM_1^T = I \quad \text{or} \quad FA^T E^T = I$$
$$A^{-1} = F^T E.$$

Here F and E are products of elementary matrices.

7.2.1 EXAMPLE. In the following example, we choose $p = 101$ and $(2R - 1) = 9$. In order to obtain E and F matrices we use the augmented matrix $[A \,|\, I]$ and $[M_1^T \,|\, I]$.

$$(A \,|\, I) = \begin{bmatrix} x & x & x-1 & 1 & 0 & 0 \\ x^2+x & x^2+2x & x^2-1 & 0 & 1 & 0 \\ 2x^2-2x & x^2-2x & 2x^2-3x+2 & 0 & 0 & 1 \end{bmatrix}$$

$$\begin{bmatrix} (1; 0) & (1; 0) & (100, 1; -1) & (1; -1) & (0; 0) & (0; 0) \\ (0; 0) & (0, 1; 0) & (0; 0) & (100, 100; 0) & (1; 0) & (0; 0) \\ (0; 0) & (0, 0, 100; 0) & (0, 1; 0) & (2, 99; 0) & (0; 0) & (1; 0) \end{bmatrix}$$

$$\begin{bmatrix} (1; 0) & (1; 0) & (100, 1; -1) & (1; -1) & (0; 0) & (0; 0) \\ (0; 0) & (1; 0) & (0; 0) & (100, 100; -1) & (1; -1) & (0; 0) \\ (0; 0) & (0; 0) & (0, 1; 0) & (2, 98, 100; 0) & (0, 1; 0) & (1; 0) \end{bmatrix}$$

$$\begin{bmatrix} (1;0) & (1;0) & (100,1;-1) & (1;-1) & (0;0) & (0;0) \\ (0;0) & (1;0) & (0;0) & (100,100;-1) & (1;-1) & (0;0) \\ (0;0) & (0;0) & (1;0) & (2,98,100;-1) & (1;0) & (1;-1) \end{bmatrix}.$$

Thus

$$M_1 = \begin{bmatrix} (1;0) & (1;0) & (100,1;-1) \\ (0;0) & (1;0) & (0;0) \\ (0;0) & (0;0) & (1;0) \end{bmatrix}$$

$$E = \begin{bmatrix} (1;-1) & (0;0) & (0;0) \\ (100,100;-1) & (1;-1) & (0;0) \\ (2,98,100;-1) & (1;0) & (1;-1) \end{bmatrix}.$$

We augment M_1^T and obtain $(M_1^T \mid I)$

$$\begin{bmatrix} (1;0) & (0;0) & (0;0) & (1;0) & (0;0) & (0;0) \\ (1;0) & (1;0) & (0;0) & (0;0) & (1;0) & (0;0) \\ (100,1;-1) & (0;0) & (1;0) & (0;0) & (0;0) & (1;0) \end{bmatrix}$$

$$\begin{bmatrix} (1;0) & (0;0) & (0;0) & (1;0) & (0;0) & (0;0) \\ (0;0) & (1;0) & (0;0) & (100;0) & (1;0) & (0;0) \\ (0;0) & (0;0) & (1;0) & (1,100;-1) & (0;0) & (1;0) \end{bmatrix}.$$

Thus

$$F = \begin{bmatrix} (1;0) & (0;0) & (0;0) \\ (100;0) & (1;0) & (0;0) \\ (1,100;-1) & (0;0) & (1;0) \end{bmatrix}; \quad U = \begin{bmatrix} (1;0) & (0;0) & (0;0) \\ (0;0) & (1;0) & (0;0) \\ (0;0) & (0;0) & (1;0) \end{bmatrix}.$$

Thus

$$F^T U E = A^{-1} = \begin{bmatrix} (2,98,3,1;-2) & (100;0) & (1,100;-2) \\ (100,100;-1) & (1;-1) & (0;0) \\ (2,98,100;-1) & (1;0) & (1;-1) \end{bmatrix}$$

or

$$A^{-1} = \begin{bmatrix} \dfrac{(2-3x+3x^2+x^3)}{x^2} & -1 & \dfrac{(1-x)}{x^2} \\[3mm] -\dfrac{(1+x)}{x} & \dfrac{1}{x} & 0 \\[3mm] \dfrac{(2-3x-x^2)}{x} & 1 & \dfrac{1}{x} \end{bmatrix}$$

7.3 Generalized inverses

If a matrix is rectangular or if a square matrix is singular, it does not have an inverse in the usual sense, but under certain conditions there exists, a generalized inverse (called g-inverse) (Gregory and Krishnamurthy [1984], Chapter III; Rao and Mitra [1971], Bhaskara Rao [1981]).

If A is an $m \times n$ matrix over a field F and G is a g-inverse of A, then G is an $n \times m$ matrix defined as follows.

7.3.1 *Various g-inverses.* Consider the matrix equations

(a) $AGA = A$
(b) $GAG = G$
(c) $(AG)^T = AG$
(d) $(GA)^T = GA$.

The matrix G is called

(i)   a g-inverse of A, denoted by A^- if (a) holds;
(ii)   a reflexive g-inverse of A, denoted by A_R^- if both (a) and (b) hold;
(iii)   a least squares g-inverse of A denoted by A_L^- if both (a) and (c) hold;
(iv)   a minimum norm g-inverse of A denoted by A_M^- if both (a) and (d) hold; and
(v)   the Moore–Penrose inverse of A, denoted by A^+, if (a), (b), (c), and (d) all hold.

Note that when A is a square nonsingular matrix, a g-inverse reduces to the ordinary inverse A^{-1}.

7.3.2 *g-inverse of polynomial matrices over $F(x)$.* If a matrix $A(x) \in F(x)$ then we can define the various g-inverses as in Section 7.3.1. The algorithms described in Gregory and Krishnamurthy [1984] can be used for this purpose, using the polynomial Hensel Codes, by a proper choice of R and p.

We first obtain $M = (AA^T)^2$ and find its reflexive inverse M_R^- using elementary row (column) operations:

$$EM = M_1$$

$$FM_1^T = U = \begin{bmatrix} I_r & 0 \\ 0 & 0 \end{bmatrix}$$

(E and F are products of elementary matrices over $F(x)$). Thus $M_R^- = F^T U E$.

We then obtain A^+, the Moore–Penrose inverse of A, using

$$A^+ = A^T M_R^- (AA^T).$$

The following simple example demonstrates this method.

7.3.3 EXAMPLE. Let

$$M = \begin{bmatrix} 1+x & 1+x \\ 1+x & 1+x \end{bmatrix} \in F_7(x);$$

find M_R^-. We choose $2R - 1 = 5$.

$$E_1 M = \begin{bmatrix} 1+6x+x^2+6x^3+x^4 & 0 \\ 0 & 1 \end{bmatrix} \begin{bmatrix} 1+x & 1+x \\ 1+x & 1+x \end{bmatrix} = \begin{bmatrix} 1 & 1 \\ 1+x & 1+x \end{bmatrix}$$

$$E_2 E_1 M = \begin{bmatrix} 1 & 0 \\ 0 & 1+6x+x^2+6x^3+x^4 \end{bmatrix} \begin{bmatrix} 1 & 1 \\ 1+x & 1+x \end{bmatrix} = \begin{bmatrix} 1 & 1 \\ 1 & 1 \end{bmatrix}$$

$$E_3 E_2 E_1 M = \begin{bmatrix} 1 & 0 \\ -1 & 0 \end{bmatrix} \begin{bmatrix} 1 & 1 \\ 1 & 1 \end{bmatrix} = \begin{bmatrix} 1 & 1 \\ 0 & 0 \end{bmatrix} = M_1$$

$$F M_1^T = \begin{bmatrix} 1 & 0 \\ -1 & 1 \end{bmatrix} \begin{bmatrix} 1 & 0 \\ 1 & 0 \end{bmatrix} = \begin{bmatrix} 1 & 0 \\ 0 & 0 \end{bmatrix} = U.$$

Thus

$$M_R^- = \begin{bmatrix} 1 & -1 \\ 0 & 1 \end{bmatrix} \begin{bmatrix} 1 & 0 \\ 0 & 0 \end{bmatrix} \begin{bmatrix} 1 & 0 \\ -1 & 1 \end{bmatrix} \begin{bmatrix} 1+6x+x^2+6x^3+x^4 & 0 \\ 0 & 1+6x+x^2+6x^3+x^4 \end{bmatrix}$$

$$= (1+6x+x^2+6x^3+x^4) \begin{bmatrix} 1 & 0 \\ 0 & 0 \end{bmatrix}$$

which when converted to $P(2/2, F_7(x))$ gives:

$$M_R^- = \frac{1}{1+x} \begin{bmatrix} 1 & 0 \\ 0 & 0 \end{bmatrix}.$$

7.3.4 Remark. It is clear that by a proper choice of degree R and modulus p we can use Hensel Code arithmetic for computing the g-inverse of matrices over $\mathbb{Q}(x)$.

7.4 Generalized inverse by evaluation–interpolation

The techniques described in Chapters II and III are applicable to find the generalized inverse of a polynomial matrix by evaluation–interpolation. We will not discuss this topic any further, as these are straightforward extensions of the methods already described.

However, in using the modular arithmetic, it may be required to take some precautionary measures to avoid failures. These are mentioned in Section 5, Chapter III, Gregory and Krishnamurthy [1984].

8 Hensel–Newton–Schultz Iterative Matrix Inversion

In Chapter V of Gregory and Krishnamurthy [1984] we described how the Newton–Schultz method can be extended to invert matrices over $\mathbb{Q}$ using p-adic computation (or Hensel Codes). We now describe how this method can further be extended to invert matrices whose elements belong to $F_p(x)$ or $\mathbb{Q}(x)$.

Let us, first, assume that $A = (a_{ij}(x))$ is a nonsingular $(n \times n)$ matrix whose elements are polynomials over $\mathbb{I}_p$. Let us also assume that the elements of A^{-1} belong to $P(R - 1/R - 1, F_p(x))$.

The Hensel–Newton–Schultz method first constructs $A^{-1}(\bmod x)$ by any direct method:

(8.1) $A^{-1}(\bmod x) = B_{2^0} = B_1$ (we assume $A^{-1} \bmod x$ exists).

Then B_1 is used as the initial or starting approximation and the iteration

(8.2) $$B_{2^i} \equiv B_{2^{i-1}}[2I - AB_{2^{i-1}}] \bmod x^{2^i}   (i \geqslant 1)$$

is performed until x^{2^i} is large enough to accommodate the inverse matrix elements belonging to the set of Padé rationals $P(R - 1/R - 1, F_p(x))$.

Then the matrix B_{2^i} is the polynomial Hensel Code of $A^{-1}(x)$ or $H(p, 2^i, A^{-1}(x))$, so that the rational polynomial elements of A^{-1} can be obtained using the conversion methods described in Section 6 of this chapter.

The following theorem ensures the validity of this method.

8.1 **Theorem.** *Let A be an $(n \times n)$ nonsingular matrix $\in F_p(x)$ and let $A \bmod x$ be nonsingular with $A^{-1} = B_1 \bmod x$. Then the iteration $B_{2^i} = B_{2^{i-1}}[2I - AB_{2^{i-1}}] \bmod x^{2^i}$, $i \geqslant 1$ generates a polynomial matrix sequence $\{B_{2^i}\}$, $i \geqslant 1$ in $\mathbb{I}_p[[x]]$ such that*

$$AB_{2^i} \equiv I \bmod x^{2^i},$$

where I is the identity matrix of order $n \times n$. In other words B_{2^i} is the Hensel Code of $A^{-1} \bmod x^{2^i}$.

PROOF. We show that the sequence $\{B_{2^i}\}$ can be generated recursively and then prove by induction that for $i \geqslant 0$

(8.3) $$AB_{2^i} \bmod x^{2^i} = I.$$

The first member of this sequence, from (8.1) is

$$B_1 = A^{-1} \bmod x.$$

Hence by (8.1)

(8.4) $$AB_1 \bmod x = I.$$

Then using (8.2)

(8.5) $$AB_{2^i} \bmod x^{2^i} = [A(B_{2^{i-1}}(2I - AB_{2^{i-1}}) \bmod x^{2^i})] \bmod x^{2^i}$$
$$= AB_{2^{i-1}}(2I - AB_{2^{i-1}}) \bmod x^{2^i}.$$

The inductive hypothesis (8.3) means

(8.6) $AB_{2^{i-1}} = I + x^{2^{i-1}} E_{i-1},$

where E_{i-1} is a residual matrix $\in F_p(x)$.

Using (8.6) in (8.5) we obtain

$$AB_{2^i} \bmod x^{2^i} = (I + x^{2^{i-1}} E_{i-1})[2I - (I + x^{2^{i-1}} E_{i-1})] \bmod x^{2^i}$$
$$= [I + x^{2^{i-1}} E_{i-1}][I - x^{2^{i-1}} E_{i-1}] \bmod x^{2^i}$$
$$= I.$$

Since by construction, the theorem holds for $i = 0$, and since the assumption that the result is true for $(i-1)$ implies that it is true for i, it is true for all $i \geqslant 0$. $\square$

8.1.1 *Number of iterations*. The above scheme is quadratically convergent. Therefore if we perform i iterations, where $2^i \geqslant (2R-1)$, we can reconstruct the elements of $A^{-1}(x)$ belonging to $P(R-1/R-1, F_p(x))$ uniquely, using the Padé or Euclidean method.

8.1.2 *Higher order convergence*. It is possible to use the higher order convergent schemes (as in Section 5.5.9). For example, the iteration

$$B_{q^k} = B_{q^{k-1}}[I + D_{k-1}(I + D_{k-1}(I + \cdots) \cdots)] \bmod x^{q^k},$$

where there are $(q-1)$ terms in the nested expression and where

$$D_{k-1} = [I - AB_{q^{k-1}}] \bmod x^{q^k}$$

gives qth order convergence.

8.2 EXAMPLE. Let

$$A = \begin{bmatrix} x^2 & x^2 + 2x + 2 \\ x + 2 & x + 2 \end{bmatrix},$$

where $a_{ij}(x) \in F_3(x)$. The maximal degree $(R-1)$ of the determinant is 3. Hence we need to iterate i times such that

$$2^i \geqslant (2R-1) = 7 \quad \text{or } i \geqslant 3.$$

We obtain by Hensel coding

$$A \bmod x = \begin{bmatrix} (0;0) & (2;0) \\ (2;0) & (2;0) \end{bmatrix}$$

and

$$B_{2^0} = B_1 = A^{-1} \bmod x = \begin{bmatrix} (1;0) & (2;0) \\ (2;0) & (0;0) \end{bmatrix}$$

$$B_{2^1} = B_2 = \begin{bmatrix} (1,2;0) & (2,2;0) \\ (2,1;0) & (0,0;0) \end{bmatrix}$$

$$B_{2^2} = B_4 = \begin{bmatrix} (1,2,1,2;0) & (2,2,0,2;0) \\ (2,1,2,1;0) & (0,0,2;0) \end{bmatrix}$$

and

$$B_{2^3} = B_8 = \begin{bmatrix} (1,2,1,2,1,2,1,2;0) & (2,2,0,2,0,2,0,2;0) \\ (2,1,2,1,2,1,2,1;0) & (0,0,2,0,2,0,2,0;0) \end{bmatrix}.$$

We stop the iteration at this step.

On conversion (Section 6) to rationals, we obtain

$$A^{-1} = \begin{bmatrix} 1/(x+1) & (x^2+2x+2)/(2x^2+1) \\ 2/(x+1) & (2x^2)/(2x^2+1) \end{bmatrix}.$$

8.3 Inversion of matrices in $\mathbb{Q}(x)$

In order to use the above method for inverting matrices whose entries belong to $\mathbb{Q}(x)$, we need to choose a suitable prime p or a product of s primes $M = \prod_{i=1}^{s} p_i$ such that

$$p \geqslant 2RN^2 + 1$$

or

$$M \geqslant 2RN^2 + 1,$$

where $(R-1)$ is the maximum degree of the numerator and denominator polynomials in the inverse matrix and N is the maximal coefficient size in the numerator/denominator polynomials. The estimates for R and N, and hence for p (or M) can be obtained by the method described in Chapter II.

We then choose the number of iterations i such that

$$2^i \geqslant (2R-1).$$

The resulting iterate B_{2^i} which is the polynomial Hensel Code A^{-1} is converted back into the rational form as described earlier.

8.3.1 EXAMPLE. Let

$$A = \begin{bmatrix} 3+x & 1+x \\ 2+x & 2+x \end{bmatrix},$$

where $a_{ij} \in \mathbb{Q}(x)$.

Let us assume $N \leqslant 4$ and $(R-1) = 1$ or $R = 2$. Thus we choose $p = 101 > 2.2.16$; we need to iterate i times such that

$$2^i \geqslant (2R-1) = 3$$

or

$$i = 2.$$

The iterations are as follows:

$$B_1 = \begin{bmatrix} 3 & 1 \\ 2 & 2 \end{bmatrix}^{-1} = \begin{bmatrix} 51 & 25 \\ -51 & 26 \end{bmatrix}$$

$$AB_1 = \begin{bmatrix} 3+x & 1+x \\ 2+x & 2+x \end{bmatrix} \begin{bmatrix} 51 & 25 \\ -51 & 26 \end{bmatrix} = \begin{bmatrix} 1 & 51x \\ 0 & 1+51x \end{bmatrix}$$

$$(2I - AB_1) = \begin{bmatrix} 1 & -51x \\ 0 & 1-51x \end{bmatrix}$$

$$B_2 = B_1(2I - AB_1) = \begin{bmatrix} 51 & 25+63x \\ -51 & 26+63x \end{bmatrix}$$

$$AB_2 = \begin{bmatrix} 1 & 25x^2 \\ 0 & 1+25x^2 \end{bmatrix}$$

$$(2I - AB_2) = \begin{bmatrix} 1 & -25x^2 \\ 0 & 1-25x^2 \end{bmatrix}$$

$$B_{2^2} = B_4 = \begin{bmatrix} 51 & (41x^3+19x^2+63x+25) \\ 50 & (41x^3+19x^2+63x+26) \end{bmatrix}$$

which in Hensel code form is

$$B_4 = \begin{bmatrix} (51;0) & (26,63,19,41;0) \\ (50;0) & (26,63,19,41;0) \end{bmatrix}.$$

On conversion to $P(1/1, F_{101}(x))$ (Section 6) B_4 becomes

$$|A^{-1}|_{101} = \begin{bmatrix} 51 & (25+25x)/(1+51x) \\ 50 & (26+76x)/(1+51x) \end{bmatrix}.$$

Each one of the coefficients in $|A^{-1}|_{101}$ on conversion using the Algorithm 4.8.2, gives,

$$A^{-1} = \begin{bmatrix} 1/2 & -\dfrac{(1/4+1/4x)}{(1+1/2x)} \\ -1/2 & \dfrac{(3/4+1/4x)}{(1+1/2x)} \end{bmatrix}$$

$$= \begin{bmatrix} 1/2 & -\dfrac{(1+x)}{4+2x} \\ -1/2 & \dfrac{(3+x)}{4+2x} \end{bmatrix} \in P(1/1, 4, x).$$

8.3.2 **Remarks.** (i) When the coefficient size N is large we can choose a set of s distinct primes $p_1, p_2, \ldots, p_s$ (or their powers $p_i^{r_i}$) such that

$$M = \prod_{i=1}^{s} p_i^{r_i} \geq 2N^2R + 1.$$

The computation is carried out independently (and in parallel, if speed is needed) with each p_i. The resulting entries of $|A^{-1}|_{p_i}$ are then combined using the Chinese remainder theorem to obtain $|A^{-1}|_M$; these would then be the Hensel Codes $H(M, r, \alpha(x))$. Using the methods of Section 6, these are converted to numerator and denominator polynomials in $\mathbb{I}_M[x]$. The coefficients in the numerator and denominator polynomials which belong to $\mathbb{I}_M$ are then converted to rationals in P_N. (See Example 8.3.3 below).

(ii) If $A \bmod x$ happens to be singular, it may be possible to introduce a suitable change of variable, say, $x + a = x$, so that the resulting matrix is invertible $(\bmod\ x)$.

(iii) It is possible to extend the above method to compute g-inverses (see Section 8.4).

8.3.3 EXAMPLE. Let

$$A = \begin{bmatrix} 3+x & 1+x \\ 2+x & 2+x \end{bmatrix},$$

where $a_{ij} \in \mathbb{Q}(x)$.

We now illustrate the use of three different primes 3, 5, 7. Here,

$$|A|_3 = \begin{bmatrix} x & (1+x) \\ (2+x) & (2+x) \end{bmatrix}; \quad |A^{-1}|_3 = \begin{bmatrix} (2;0) & (2,1,1,1;0) \\ (1;0) & (0,1,1,1;0) \end{bmatrix}$$

$$|A|_5 = \begin{bmatrix} (3+x) & (1+x) \\ (2+x) & (2+x) \end{bmatrix}; \quad |A^{-1}|_5 = \begin{bmatrix} (3;0) & (1,3,1,2;0) \\ (2;0) & (2,3,1,2;0) \end{bmatrix}$$

$$|A|_7 = \begin{bmatrix} (3+x) & (2+x) \\ (2+x) & (2+x) \end{bmatrix}; \quad |A^{-1}|_7 = \begin{bmatrix} (4;0) & (5,6,4,5;0) \\ (3;0) & (6,6,4,5;0) \end{bmatrix}.$$

Using Chinese remainder theorem and combining, we obtain:

$$|A^{-1}|_{105} = \begin{bmatrix} (53;0) & (26,13,46,82;0) \\ (52;0) & (27,13,46,82;0) \end{bmatrix}$$

which on conversion to $P(1/1, F_{105}(x))$ results in

$$|A^{-1}|_{105} = \begin{bmatrix} 53 & (26+26x)/(1+53x) \\ 52 & (27+79x)/(1+53x) \end{bmatrix}.$$

The coefficients of the entries in $|A^{-1}|_{105}$ are now converted to rationals using the Euclidean algorithm. Thus

$$A^{-1} = \begin{bmatrix} 1/2 & \dfrac{-(1/4+1/4x)}{(1+1/2x)} \\ -1/2 & \dfrac{(3/4+1/4x)}{(1+1/2x)} \end{bmatrix}$$

$$= \begin{bmatrix} 1/2 & \dfrac{-(1+x)}{(4+2x)} \\ -1/2 & \dfrac{(3+x)}{(4+2x)} \end{bmatrix} \in P(1/1, 4, x)$$

8.4 Generalized inversion by iteration

The Hensel–Newton–Schultz method can be generalized to find the g-inverse of a matrix over $F_p(x)$. Let A be an $(m \times n)$ matrix and let $M = (AA^T)^2$. Then we have $M_R^- = [(AA^T)^2]_R^-$, where $M_R^- = (\mu_{ij})$, denotes a reflexive g-inverse of M. Let us assume that $\mu_{ij} \in P(R - 1/R - 1, F_p(x))$. Also let k be an integer such that

$$2^k \geqslant 2R - 1.$$

If we map the elements μ_{ij} into $F_p(x)$ we have

$$M_R^- \bmod x^{2^k} = B_{2^k} = B_{2^0} + (B_{2^1} - B_{2^0}) + \cdots + (B_{2^k} - B_{2^{k-1}});$$

where

$$B_{2^0} = M_R^- \bmod x$$
$$B_{2^1} = M_R^- \bmod x^2$$
$$\vdots$$
$$B_{2^k} = M_R^- \bmod x^{2^k}.$$

Consequently when M_R^- exists, B_{2^i} corresponds to the truncated formal matrix power series representation of M_R^- or the Hensel Code of M_R^-. To realize M_R^-, it must be possible

(i)   to compute $M_R^- \bmod x = B_1$, and for $1 < i \leqslant k$
(ii) to compute $M_R^- \bmod x^{2^i} = B_{2^i}$ given $B_{2^{i-1}}$.

We now generalize the iteration (8.2), for the generalized matrix inversion.

8.4.1 Algorithm

Step 1. Construct

(8.7) $$M_R^- \bmod x = B_1$$

using the direct method (Section 7.3.2). Thus we have

(8.8) $$B_1 = B_1 M B_1 \bmod x$$

and

$$M = M B_1 M \bmod x.$$

Step 2. For $i = 1, 2, \ldots, k$ compute the sequence

(8.9) $$B_{2^i} = B_{2^{i-1}}(2I - M B_{2^{i-1}}) \bmod x^{2^i}.$$

Step 3. At this point $M_R^- = B_{2^k}$;

(8.10) $$A^+ = A^T M_R^-(AA^T).$$

This algorithm rests on the following theorem.

8.4.2 **Theorem.** *Let M be an $(m \times m)$ matrix $\in F_p(x)$ and let M mod x possess a reflexive generalized inverse M_R^- mod x, satisfying $M_R^- M M_R^- = M_R^-$ mod x and $M M_R^- M = M$ mod x. Also let M_R^- mod $x = B_1 = B_{2^0}$. Then the iteration (8.9)*

$$(8.9) \qquad B_{2^i} = B_{2^{i-1}}[2I - MB_{2^{i-1}}] \bmod x^{2^i}, \qquad i \geq 1$$

generates a polynomial matrix sequence $\{B_{2^i}\}$, $i \geq 1$, in $\mathbb{I}_p[[x]]$ such that

$$B_{2^i} M B_{2^i} = B_{2^i} \bmod x^{2^i}$$

$$M B_{2^i} M = M \bmod x^{2^i}.$$

PROOF. The proof is analogous to Theorem 8.1. We must, however, note here that M could be singular or nonsingular; in the development of the proof, therefore, we need to modify the inductive hypothesis (8.6) to suit this requirement.

It is required to prove that when M_R^- exists and $2^k \geq 2R - 1$, we can compute M_R^- using the iteration (8.9) and satisfying the properties:

$$(8.11) \qquad\qquad MB_{2^i}M \bmod x^{2^i} = M$$

$$(8.12) \qquad\qquad B_{2^i}MB_{2^i} \bmod x^{2^i} = B_{2^i}$$

for $i = 1, 2, \ldots, k$.

We prove this by induction. For $i = 0$, this is true by construction:

$$(8.13) \qquad\qquad MB_1M \bmod x = M \bmod x$$

$$(8.14) \qquad\qquad B_1MB_1 \bmod x = B_1 \bmod x.$$

First, let us rewrite the equality:

$$
\begin{aligned}
MB_1M &= (MB_1)M \\
&= [I + MB_1 - I]M \\
(8.15) \qquad\qquad &= \left[I + x\left[\frac{MB_1 - I}{x}\right] \right]M \\
&= [I + xK(x)]M,
\end{aligned}
$$

where $K(x) \in F_p(x)$ and similarly

$$(8.16) \qquad\qquad B_1MB_1 = B_1(I + xR(x))$$

or

$$(8.17) \qquad\qquad B_1MB_1 = (I + xS(x))B_1,$$

where $R(x),\ S(x) \in F_p(x)$.

The equations (8.15), (8.16), and (8.17) suggest that we can write the inductive hypothesis for the step $(i - 1)$ as:

$$(8.18) \qquad\qquad MB_{2^{i-1}}M = [I + x^{2^{i-1}}K(x)]M$$

$$(8.19) \qquad\qquad B_{2^{i-1}}MB_{2^{i-1}} = B_{2^{i-1}}[I + x^{2^{i-1}}R(x)]$$

$$(8.20) \qquad\qquad B_{2^{i-1}}MB_{2^{i-1}} = [I + x^{2^{i-1}}S(x)]B_{2^{i-1}}$$

to satisfy the requirements:

(8.21) $$MB_{2^{i-1}}M \bmod x^{2^{i-1}} = M$$

(8.22) $$B_{2^{i-1}}MB_{2^{i-1}} \bmod x^{2^{i-1}} = B_{2^{i-1}}.$$

Now to prove the theorem we assume the truth of (8.18), (8.19), and (8.20) at step $(i-1)$ and show that the iteration (8.9) results in the truth of (8.18), (8.19), and (8.20) for the succeeding step i [or satisfy (8.11) and (8.12)].

We have by (8.9)

$$MB_{2^i}M = M[B_{2^{i-1}}(2I - MB_{2^{i-1}})]M$$
$$= [2MB_{2^{i-1}}M - MB_{2^{i-1}}MB_{2^{i-1}}M].$$

Using (8.18) this simplifies to

$$MB_{2^i}M = [2(I + x^{2^{i-1}}K)M - (I + x^{2^{i-1}}K)(I + x^{2^{i-1}}K)M]$$
$$= (I + x^{2^{i-1}}K)(2M - M - x^{2^{i-1}}KM)$$
$$= (I + x^{2^{i-1}}K)(I - x^{2^{i-1}}K)M$$
$$= (I - x^{2^i}K^2)M$$
$$= (I + x^{2^i}T)M,$$

where $-K^2 = T$, as was to be proved.

Similarly by (8.9)

$$B_{2^i}MB_{2^i} = B_{2^{i-1}}[2I - MB_{2^{i-1}}]MB_{2^{i-1}}[2I - MB_{2^{i-1}}]$$
$$= [2B_{2^{i-1}} - B_{2^{i-1}}MB_{2^{i-1}}]M[2B_{2^{i-1}} - B_{2^{i-1}}MB_{2^{i-1}}].$$

Using (8.19) and (8.20) this simplifies to

$$[2B_{2^{i-1}} - B_{2^{i-1}} - x^{2^{i-1}}SB_{2^{i-1}}]M[2B_{2^{i-1}} - B_{2^{i-1}} - x^{2^{i-1}}B_{2^{i-1}}R]$$
$$= [I - x^{2^{i-1}}S]B_{2^{i-1}}[I + x^{2^{i-1}}R][I - x^{2^{i-1}}R].$$

This reduces to

$$[I - x^{2^{i-1}}S]B_{2^{i-1}}[I - x^{2^i}R^2] = [2I - (I + x^{2^{i-1}}S)]B_{2^{i-1}}[I - x^{2^i}R^2]$$

which on using (8.20) gives

$$[2B_{2^{i-1}} - B_{2^{i-1}}MB_{2^{i-1}}][I - x^{2^i}R^2]$$
$$= B_{2^{i-1}}[2I - MB_{2^{i-1}}][I - x^{2^i}R^2] = B_{2^i}[I - x^{2^i}R^2] = B_{2^i}[I + x^{2^i}V],$$

where $V = -R^2$, as was to be proved. $\square$

8.4.3 **Remark.** The above scheme is quadratically convergent. One can naturally think of higher order convergent schemes as described in Section 8.1.2.

8.4.4 **Remark.** In order to extend the above algorithm for polynomial matrices with rational coefficients we need to make proper choice of N and R as explained before (Section 8.3).

8.4.5 EXAMPLE. Let

$$M = \begin{bmatrix} 1+x & 1+x \\ 1+x & 1+x \end{bmatrix} \in F_3(x); \quad \text{find } M_R^-.$$

We choose $(2R - 1) = 3$; $R - 1 = 1$; thus $k \geq 2$. We have

$$M_R^- \bmod x = \begin{bmatrix} 1 & 1 \\ 1 & 1 \end{bmatrix}_R^- = \begin{bmatrix} 1 & 0 \\ 0 & 0 \end{bmatrix} = B_1$$

(see Example 7.3.3).

We then obtain using (8.9)

$$B_2 = B_1[2I - MB_1] \bmod x^2$$

$$= \begin{bmatrix} 1 & 0 \\ 0 & 0 \end{bmatrix} \left[\begin{pmatrix} 2 & 0 \\ 0 & 2 \end{pmatrix} - \begin{pmatrix} 1+x & 1+x \\ 1+x & 1+x \end{pmatrix} \begin{pmatrix} 1 & 0 \\ 0 & 0 \end{pmatrix} \right] \bmod x^2$$

$$= \begin{bmatrix} 1-x & 0 \\ 0 & 0 \end{bmatrix};$$

$$B_4 = B_2[2I - MB_2] \bmod x^4$$

$$= \begin{bmatrix} 1-x & 0 \\ 0 & 0 \end{bmatrix} \left[\begin{pmatrix} 2 & 0 \\ 0 & 2 \end{pmatrix} - \begin{pmatrix} 1+x & 1+x \\ 1+x & 1+x \end{pmatrix} \begin{pmatrix} 1-x & 0 \\ 0 & 0 \end{pmatrix} \right] \bmod x^4$$

$$= \begin{bmatrix} 1-x & 0 \\ 0 & 0 \end{bmatrix} \begin{bmatrix} 1+x^2 & 0 \\ -1+x^2 & 0 \end{bmatrix} = \begin{bmatrix} 1-x+x^2-x^3 & 0 \\ 0 & 0 \end{bmatrix}.$$

We stop the iteration here. The result on conversion to Padé rational polynomial is given by:

$$M_R^- = \begin{bmatrix} 1/1+x & 0 \\ 0 & 0 \end{bmatrix}$$

(see Example 7.3.3).

8.4.6 **Remark.** When an estimate for R and hence for k is not available, the "for" loop in step 2 of algorithm can be replaced by a "while" loop thus: while

$$\text{Padé } B_{2^{i-1}} \neq \text{Padé } B_{2^{i-2}} \text{ do}$$

$$B_{2^i} \leftarrow B_{2^{i-1}}(2I - MB_{2^{i-1}}) \bmod x^{2^i};$$

here the Padé function (Sections 4.8, 6.2) associates with each element in the matrices $B_{2^{i-1}}$ and $B_{2^{i-2}}$, the corresponding order $R - 1/R - 1$ rational polynomials; for termination we check whether the same rational polynomials are produced at steps $i-1$ and $i-2$; if yes, then $M_R^- = B_{2^{i-1}}$ is determined unambiguously.

8.4.7 *Complexity considerations.* The complexity estimate for the operations involved in the above iterative scheme can be obtained as follows: To start with we assume M is an $m \times m$ matrix whose elements are polynomials with integral coefficients with μ as the maximal coefficient and $(d-1)$ as the maximal degree. Then the rational polynomial elements of the inverse (or g-inverse) can have a coefficient size

$$E \leqslant m!\, \mu^m d^{m-1}$$

and degree size

$$D \leqslant m(d-1)$$

(see Chapter II, Sections 2.6.8 and 2.6.9) for the numerator and denominator polynomials.

At each iterative step i we need to multiply two polynomial matrices whose elements are of degrees at most $(2^i - 1)$. Thus if we require a (D/D) Padé approximant as the final answer, we should have a Hensel code of length $(2D+1)$ at the last step of iteration. Also the choice of prime p (or s primes) should be such that $\prod_{i=1}^{s} p_i \geqslant p \geqslant E$. The number of iterations k should therefore be such that

$$2^k \geqslant (2D+1).$$

Thus for k steps we need

$$2m^3 \sum_{i=0}^{k} (2^i)^2 = O(4^k m^3)$$

coefficient multiplications of precision $\log_2 p$ bits or $s \log_2 p_i$ bits when s primes are used.

The final conversion from Hensel Code to Padé rationals require $O(m^2 D \log^2 D)$ multiplications (Brent, Gustavson, and Yun [1980]); neglecting this, the order of complexity of the Hensel–Newton–Schultz (HNS) method is $O(4^k m^3)$ multiplications of $\log_2 p$ precision operands.

8.5 Recursive property

In Sections 3 and 4.6 of this chapter, we mentioned that the concept of isomorphism plays an important role in computation. We also explained how the arithmetic of rational numbers can be realized by using the concept of explicit isomorphism between the algebra of rational numbers and the ring of integers modulo a prime power; also, we described how these ideas can be extended to the algebra of rational polynomials over a finite field. In both these cases, we essentially solve the image problem in a simpler domain I_p or $(I_p[x])$ rather than solving the problem in the more complex (original) domain Q (or $Q(x)$), using the isomorphism among the domains. This is done in two parts: we first carry out the computation in the isomorphic domain; then we use a semantic valuation function which associates with each element in the image domain, a corresponding element in the original domain. The concept of domain isomorphism and recursive computations in reflexive domains are discussed in detail by Scott and Strachey [1971]; see also Stoy [1977], Tourlakis [1984], Tennent [1981], de Bakker [1980], Kfoury, Moll, and Arbib [1982]. With this in mind we now outline some of the properties of the Hensel–Newton–Schultz (HNS) method described in Section 8.

The HNS method is branded "iterative" for historical reasons, since it mimics the well-known Newton's iterative method in numerical analysis. In the Newton's method, over the reals, the iteration drives the result to a constant (unique fixed point) of the system based on Banach contraction mapping principle in a metric space. The HNS method, however, is essentially an inductive or recursive method (over the domains of matrices whose elements belong to the Hensel field)—leading to Kleene's least fixed point over an ordered set (lattice); here, the recursive steps progressively expand the words (string of symbols that denote the power series expansion of each element in the matrix) in the system, such that an input word from a previous step is contained in the output word from the following computational step. Such a recursive method is initialized with the weakest solution; this is then improved monotonically towards better and better solutions, and eventually to the well-defined solution and overdefined solutions (Scott and Strachey [1971]). We therefore look for a minimal or least fixed point; step i for which $2^i \geqslant 2R - 1$ is the least fixed point of this recursively defined matrix function. This is because (a) it is the least i for which we obtain the well-defined solution that contains all the underdefined solutions and (b) it is the least i for which the solution is well-defined and is contained in all other succeeding overdefined solutions. Thus all these solutions form a complete lattice (de Bakker [1980], Barendregt [1984], Scott [1976]).

For a study on the connection between the least fixed point for ordered sets and a unique fixed point for metric spaces under contraction mapping; see Arbib and Manes [1982], Kelley and Namioka [1976].

8.6 Relation to diophantine approximation

We now relate the Hensel Code approach to solving the matrix inversion problems, with the well-known diophantine approximations (Mahler [1961], Le Veque [1977], Stolarsky [1974]). Diophantine approximation is concerned with the problem of best approximation of a real number (or polynomial in several variables with real coefficients) by a rational number (or rational polynomial). The continued fraction approximation belongs to this category; other closely related topics are the p-adic and power series approximants.

In all our discussions we are concerned with the finitely terminating diophantine approximation of the rational numbers (rational functions); hence the classical iterative procedure for linear problems takes the form of recursive procedure on number theoretic functions.

EXERCISES IV

1. Find the inverse transform $a(x)^{-1}$ where $a(x) = (2x+4) \in \mathbb{I}_{101}[x]$.

2. Given $H(601, 5, \alpha(x)) = (1, 300, 2, 590, 24; 0)$ find the equivalent Padé rational polynomial.

3. Find the inverse transform of $H(17, 4, \alpha(x)) = (1, 8, 7, 4; 0)$ using

   (i)   long division;
   (ii)   Euclidean algorithm;
   (iii) Newton–Hensel iteration.

4. Find the inverse of:

$$A = \begin{bmatrix} x & x+1 \\ 2x & 3x \end{bmatrix},$$

   where $a_{ij} \in F_5(x)$, using the reduction algorithm.

5. Find the inverse of:

$$\begin{bmatrix} x & x+1 & x+2 \\ x+2 & x+3 & x+4 \\ x+3 & x+4 & x+5 \end{bmatrix}$$

   with $a_{ij} \in F_{101}(x)$ using the Newton–Schultz iteration.

6. Write a program for converting any given $H(p, r, \alpha(x))$ using the Padé matrix method.

7. Find the inverse of:

$$A = \begin{bmatrix} x & x+1 & x+2 \\ x+2 & x+3 & x+4 \\ x+3 & x+4 & x+5 \end{bmatrix}$$

   over $Q(x)$.

8. Find the inverse of:

$$A = \begin{bmatrix} x & x+1 & 0 \\ x+2 & 0 & -x+1 \\ 0 & x+1 & x+2 \end{bmatrix}$$

   with $a_{ij} \in Q(x)$.

9. Convert the following matrix A into equivalent Padé rational form $P(1/1, 5, x)$:

$$A = \begin{bmatrix} 51, 0, 0; 0) & (25, 63, 19; 0) \\ 50, 0, 0; 0) & (26, 63, 19; 0) \end{bmatrix}$$

   with $a_{ij} \in F_{101}(x)$.

10. Find a generalized inverse of:

$$A = \begin{bmatrix} 1+x & 6+6x \\ 4+4x & 24+24x \end{bmatrix} \text{ over } F_{101}(x).$$

11. Find a generalized inverse of:

$$A = \begin{bmatrix} x+2 & x^2+1 \\ x+2 & x^2+1 \end{bmatrix} \text{over } F_{17}(x)$$

using Newton–Schultz method.

12. Find a generalized inverse of:

$$A = \begin{bmatrix} 1+x & 1+x & 1+x \\ 2+x & 2+x & 2+x \\ 4+4x & 4+4x & 4+4x \end{bmatrix} \text{over } F_{37}(x)$$

by any method.

13. Use the Newton–Schultz cubic convergence method to find a g-inverse of

$$\begin{bmatrix} x+1 & x+2 \\ x+1 & x+2 \end{bmatrix} \text{over } F_7(x).$$

14. Use the evaluation–interpolation method to find a g-inverse of:

$$\begin{bmatrix} x+1 & y+1 \\ x+1 & y+1 \end{bmatrix} \text{over } F_7(x,y).$$

15. Use the Fourier interpolation and evaluation to find a g-inverse of:

$$\begin{bmatrix} 3x & 2y \\ 3x & 3y \end{bmatrix} \text{over } F_{11}(x,y).$$

CHAPTER V

Matrix Computations—Euclidean and Non-Euclidean Domains

1 Introduction

So far we were essentially concerned with the inversion of matrices over fields and field of quotients constructed from the Euclidean domains (rational numbers and single variable rational functions). We now consider some of the computational problems related to matrices over the Euclidean domains (which are principal ideal rings) and general integral domains that are not Euclidean (and not principal ideal rings).

2 Matrices over Euclidean Domains

The ring of integers I and the ring of polynomials $R[x]$ in a variable x over a field are important examples of Euclidean domains.

Such matrices can be reduced only to the Smith Canonical form (SCF),

$$P\begin{bmatrix} S & 0 \\ 0 & 0 \end{bmatrix}Q,$$

where P and Q are invertible square matrices (with determinants over I or R (see Section 2.4.3, Chapter II; see also MacDuffee [1948]). The inversion and g-inversion of these matrices are based on their reduction to the SCF (Newman [1972], Gregory and Krishnamurthy [1984], Bhaskara Rao [1981]).

2.1 Matrices over I

One of the principal difficulties encountered in the computation of both the inverse and SCF of a matrix over I, is the intermediate swell of numbers and the resulting overflow. Several attempts have been made recently to solve this problem (Kannan and Bachem [1979], Kannan [1983], Alagar and Roy [1984]).

2.1.1 *Inversion of matrices*. The inversion of matrices over I can be carried out without reduction to SCF by assuming that these matrices belong to $\mathbb{Q}$. We then reduce the matrix to the $\begin{bmatrix} I_r & 0 \\ 0 & 0 \end{bmatrix}$ form (Section 2.7, Chapter I) and find A^{-1} exactly using the modular or p-adic arithmetic. If the results do not belong to I then we may say that A^{-1} does not exist over I.

2.1.2 *Smith Canonical form*. Unlike in a matrix inversion over $\mathbb{Q}$, the use of multiple modular arithmetic for the reduction of a matrix over I to SCF may not succeed. This is essentially due to the reason that the process of reduction to SCF involves the relative magnitudes of the matrix entries (and finding gcd) and there is no guarantee that the modular reduction steps obtained using the different primes will all be in one-to-one correspondence with each other and with the reduction steps based on integer arithmetic.

2.2 Matrices over $R[x]$

The computation of SCF for matrix polynomials, as in the case of matrices over I, leads to the "intermediate expression swell" and "coefficient growth" (Kannan [1983]). One might hope that the reduction to SCF could perhaps be achieved by using the evaluation–interpolation algorithm (Chapters II and III). But this procedure may not succeed, since there need not exist one-to-one correspondence in the ordering of the polynomials (based on the Euclidean valuation-degree) and in the ordering of its evaluated values (based on the Euclidean valuation-magnitude) over R or F. Thus in the process of reduction to SCF using evaluation–interpolation, the resulting matrix entries in R (or F) for the different evaluated values need not be in one-to-one correspondence with each other and with the reduction steps based on direct polynomial reduction.

In fact the analogy that exists between the evaluation–interpolation for polynomials and the multiple-modular arithmetic for integers (Section 4, Chapter I), by itself, is an indication that both these techniques might fail for reduction to SCF, in the respective cases.

2.2.1 *Inversion of matrices over* $R[x]$. As in the case of matrices over $\mathbb{I}$, we can assume that the given matrix belongs to $\mathbb{Q}(x)$ and compute the inverse using the polynomial Hensel Codes. If the result is not in $R[x]$ we say A^{-1} does not exist in $R[x]$.

3 Matrices over Non-Euclidean Domains

Two interesting examples of non-Euclidean domains are $\mathbb{I}[x]$ and $R[x_1, x_2, \ldots, x_n]$. Since the Euclidean algorithm is absent, we cannot define the SCF for matrices over $\mathbb{I}[x]$ or $R[x,y]$ or $R[x_1, \ldots, x_n]$. Some conditions for the existence of inverses and g-inverses of such matrices are available in Bhaskara Rao [1983].

In the next section, we extend the concept of Hensel Codes for the rational functions of multivariable polynomials.

4 Multivariable Polynomial Hensel Codes

We now extend by analogy all the concepts relating to the construction of Hensel Codes and arithmetic operations of single variable rational polynomials over a finite field (and rationals), to multivariable rational polynomials over a finite field (and rationals). This analogy however, does not extend to the Euclidean inverse mapping method, since the multivariable polynomials over a field are not Euclidean domains. However, we can either use the Padé method in its extended form (Chisholm [1974], [1973], [1977]) or the common denominator method (Chapter IV), for the inverse mapping.

4.1 Hensel Codes

Consider the rational polynomial in two variables x,y over a field F.

$$\frac{a(x,y)}{b(x,y)} = \alpha(x,y) = \frac{\displaystyle\sum_{i,j=0}^{m} a_{ij}x^iy^j}{\displaystyle\sum_{i,j=0}^{m} b_{ij}x^iy^j},$$

where $a_{ij}, b_{ij} \in F = (\mathbb{I}_p, +, \cdot)$, and $a_{00} \neq 0$ and $b_{00} \neq 0$.

We can expand $b(x,y)$ as a formal power series over F (Section 5.2, Chapter I) and obtain

$$C(x,y) = a(x,y) \cdot b^{-1}(x,y) = \sum_{i,j=0} c_{ij}x^iy^j + O(r),$$

where $O(r)$ denotes that the terms of the form $x^k y^{r-k}$ with degree greater than or equal to r are truncated. In other words, only those terms with degree less than or equal to $(r-1)$ are retained. We denote this truncation by $C(x,y)\,O(r)$.

The Pascal triangle is a useful aid to remember which combinations of terms are truncated for a given order r. The following scheme illustrates this.

PASCAL TECHNIQUE

r

0						1					
1					x		y				
2				x^2		xy		y^2			
3			x^3		x^2y		xy^2		y^3		
4		x^4		x^3y		x^2y^2		xy^3		y^4	
5	x^5		x^4y		x^3y^2		x^2y^3		xy^4		y^5

For more than two variables, Pascal tetrahedron and higher volume elements are used to denote the truncation $O(r)$.

By analogy with the single variable polynomial (Chapter IV), we call $c(x,y)\,O(r)$ as the Hensel Code of $\alpha(x,y)\,O(r)$, and we denote this by $H(p, r, \alpha(x,y))$.

This notion of Hensel Code of multivariable rational polynomial is useful for the arithmetic of multivariable rational polynomials over a field.

As in the case of single variable rational polynomial Hensel Code, we now describe some basic properties of the multivariable Hensel Codes, the forward and inverse mappings.

4.2 Forward mapping

We use similar definitions and notation as in Section 4.1 (Chapter IV) for denoting the various sets, by simply replacing $[x]$, $[[x]]$, (x), by the n variables $[x_1, x_2, \ldots, x_n]$, $[[x_1, x_2, \ldots, x_n]]$, $(x_1, x_2, \ldots, x_n)$ respectively.

The forward mapping is a succession of two mappings.

(i)
$$P(R-1, \ldots, R-1/R-1, \ldots, R-1, N, x_1, \ldots, x_n)$$
$$\rightarrow P(R-1, \ldots, R-1/R-1, \ldots, R-1, \hat{F}_p(x_1, x_2, \ldots, x_n)).$$

Here we map the rational polynomial $[a(x_1, \ldots, x_n)]/[b(x_1, \ldots, x_n)]$ (with integral coefficients) whose numerator and denominator degrees are at most $(R-1)$ in each variable, and the maximum coefficient size N, to the rational polynomial $\hat{F}_p(x)$ using the following rules: choose $p \geq 2R^n N^2 + 1$

and set

$$a_{ijkl\cdots} \to a_{ijkl\cdots} \quad \text{if} \quad a_{ijkl\cdots} \geqslant 0$$
$$a_{ijkl\cdots} \to p + a_{ijkl\cdots} \quad \text{if} \quad a_{ijkl\cdots} < 0;$$

Similarly

$$b_{ijkl\cdots} \to b_{ijkl\cdots} \quad \text{if} \quad b_{ijkl\cdots} \geqslant 0$$
$$b_{ijkl\cdots} \to p + b_{ijkl\cdots} \quad \text{if} \quad b_{ijkl\cdots} < 0.$$

(See Section 4.5, Chapter I.)

(ii) $\quad P(R-1, \ldots, R-1/R-1, \ldots, R-1, \hat{F}_p(x_1, \ldots, x_n)$
$$\to \hat{\mathbb{I}}_p[[x_1, \ldots, x_n]] \to \hat{\mathbb{I}}_p[x_1, \ldots, x_n].$$

Here we first map the Padé rational polynomial over a finite field to its formal power series in $\hat{\mathbb{I}}_p[[x_1, \ldots, x_n]]$ by finding $[a(x_1, \ldots, x_n) \cdot b^{-1}(x_1, \ldots, x_n)]$. This infinite series is truncated and mapped to a subset $\hat{\mathbb{I}}_p[x_1, \ldots, x_n]$ by the rule

$$\hat{\mathbb{I}}_p[x_1, \ldots, x_n] = \{[a(x_1, \ldots, x_n) \cdot b^{-1}(x_1, \ldots, x_n)]O(r):$$
$$a(x_1, \ldots, x_n)/b(x_1, \ldots, x_n) \in P(R-1, \ldots, R-1/R-1, \ldots, R-1,$$
$$\hat{F}_p(x_1, \ldots, x_n)\},$$

where $r \geqslant 2n(R-1)+1$.

In other words, $\hat{\mathbb{I}}_p[x_1, \ldots, x_n]$ denote the set of images of Padé rationals.

The following theorems ensure that the mappings described above are one-to-one and onto and they have an inverse.

4.2.1 **Theorem.** Let $\quad \alpha(x_1, x_2, \ldots, x_n) = a(x_1, x_2, \ldots, x_n) \cdot b^{-1}(x_1, x_2, \ldots, x_n)$ be a formal power series (Section 5.2, Chapter I) over

$$F = (\mathbb{I}_p, +, \cdot)$$

and let the degree of each variable in a and b be at most equal to $(R-1)$ and the truncated series $\alpha^T(x_1, x_2, \ldots, x_n)$ satisfy

$$\alpha^T - \alpha = O(r),$$

where

(4.1) $$r \geqslant 2n(R-1)+1;$$

then α^T is unique for any element in

$$P(R-1, \ldots, R-1/R-1, \ldots, R-1, \hat{F}_p(x_1, x_2, \ldots, x_n)).$$

PROOF. Similar to Theorem 4.4 of Chapter IV. $\qquad \square$

4.2.2 **Theorem** (Uniqueness of Hensel Codes). *Let N be the largest* *integer satisfying the inequality*:

(4.2) $$N \leq \sqrt{(p - 1/2R^n)};$$

then in order that $a(x_1, x_2, \ldots, x_n)/b(x_1, x_2, \ldots, x_n)$ *belonging to* $P(R - 1,$ $\ldots, R - 1/R - 1, \ldots, R - 1, N, x_1, x_2, \ldots, x_n)$ *is represented uniquely by* $\alpha^T(x_1, x_2, \ldots, x_n)$ *in* $\hat{\mathbb{I}}_p[x_1, x_2, \ldots, x_n]$, *it is sufficient that* (i) deg $a(x_1, x_2, \ldots, x_n)$ *and* deg $b(x_1, x_2, \ldots, x_n)$ *are at most* $n(R - 1)$ *(note that each variable has a degree at most* $(R - 1)$*) and* (ii) *the maximal absolute value of the coefficients* $|a_{ijkl\ldots}|$ *and* $|b_{ijkl\ldots}| \leq N$.

PROOF. The proof is similar to that for Theorem 4.5 (Chapter IV), and uses the fact that the maximal coefficient in a product of two n variable polynomials (each variable occurring with at most degree $(R - 1)$ and each coefficient is less than N) is less than or equal to $R^n N^2$ □

4.2.3 **Remarks.** (i) Note that the inequalities (4.1) and (4.2) reduce to $r \geq 2R - 1$ and $N \leq \sqrt{(p - 1)/2R}$ for the single variable polynomial $(n = 1)$ and

(ii) $$N \leq \sqrt{(p - 1)/2}$$

for a constant polynomial $(R - 1 = 0)$ which corresponds to Farey rationals of order N.

4.2.4 EXAMPLE. Let $a_{ij}, b_{ij} \in \mathbb{I}_7$; construct the Hensel code of

$$\alpha(x, y) = \frac{1 + 2x + y}{1 + x + 2y} = (1 + 2x + y)(1 + x + 2y)^{-1}.$$

We have

$$(1 + x + 2y)^{-1} = 1 + 6x + 5y + x^2 + 4y^2 + 4xy + 6x^3 + 6y^3 + x^2y + 2xy^2 + x^4$$
$$+ 2y^4 + 4xy^3 + x^3y + 3x^2y^2 + O(5).$$

Thus

$$H(7, 5, \alpha(x, y)) = 1 + x + 6y + 6x^2 + 2y^2 + 6xy + x^3 + 3y^3 + 3x^2y + 6x^4 + y^4$$
$$+ 4xy^3 + 2x^3y + x^2y^2.$$

4.3 Inverse mapping

The inverse mapping is the problem of conversion of $H(p, r, \alpha(x_1, x_2, \ldots, x_n))$ back into $\alpha(x_1, x_2, \ldots, x_n) \in P(R - 1,$ $R - 1, \ldots, R - 1/R - 1, R - 1, \ldots, R - 1, N, x_1, x_2, \ldots, x_n)$. This consists of two mappings:

$$M_1 : H(p, r, \alpha(x_1, x_2, \ldots, x_n))$$
$$\rightarrow P(R - 1, \ldots, R - 1/R - 1, \ldots, R - 1, \hat{F}_p(x_1, x_2, \ldots, x_n))$$

and
$$M_2: P(R-1, \ldots, R-1/R-1, \ldots, R-1, \hat{F}_p(x_1, x_2, \ldots, x_n)$$
$$\to P(R-1, \ldots, R-1/R-1, \ldots, R-1, N, x_1, x_2, \ldots, x_n).$$

Mapping M_1 is achieved by the solution of the Padé method (Section 6.2, Chapter IV). Since by construction of the Hensel Code it is unique, we will have adequate numbers of linear equations to uniquely solve for the denominator and numerator polynomials.

Mapping M_1 can also be achieved by the common denominator method (Section 6.4, Chapter IV). We shall not illustrate this here. Mapping M_2 uses the same procedure described in Chapter IV, Section 4.8.2, to convert each individual coefficient to its equivalent rational number.

We will illustrate the Padé method with a two variable example.

4.3.1 **Example.** Consider

$$H(7,5,\alpha(x,y)) = 1 + 6x + 5y + x^2 + 4y^2 + 4xy + 6x^3 + 6y^3 + x^2y + 2xy^2$$
$$+ x^4 + 2y^4 + 4xy^3 + x^3y + 3x^2y^2 = c(x,y).$$

We assume the form

$$\frac{a(x,y)}{b(x,y)} = \frac{a_{00} + a_{10}x + a_{01}y}{1 + b_{10}x + b_{01}y} = c(x,y).$$

By multiplying $b(x,y)$ and $c(x,y)$ and comparing the coefficients of the product with those of $a(x,y)$, we get,

$$a_{00} = 1$$
$$a_{10} = b_{10} + 6$$
$$b_{10} \cdot 6 + 1 = 0; \quad \text{hence } b_{10} = 1; \qquad a_{10} = 0, \qquad a_{01} = b_{01} + 5$$
$$b_{01} \cdot 5 + 4 = 0; \quad \text{hence } b_{01} = 2; \qquad a_{01} = 0.$$

Thus $\alpha(x,y) = 1/1 + x + 2y$.

4.3.2. **Remark.** We already mentioned that the inverse mapping of a multivariable polynomial Hensel Code to its equivalent rational polynomial cannot be realized by the conventional Extended Euclidean algorithm, since the multivariable polynomials are not Euclidean domains. Recently Buchberger and Finkler [1984] have reported that Gröbner bases, (Buchberger [1984], Buchberger, Collins, and Loos [1984]) can be a very effective tool to decode the multivariable Hensel Code. If this algorithm is efficient it will have several other applications—multivariable Padé approximation, construction and decoding of multivariable Goppa code (McEliece [1977]), and in the construction of matrix Padé approximants (Baker and Graves–Morris [1981]).

4.4 Arithmetic

The arithmetic of multivariable Hensel Codes is similar to the multivariable polynomial arithmetic (Section 5, Chapter I). It is clear that by a proper choice of p and r, the size of the results can be confined to the Padé rational chosen, so that, we can carry out multivariable rational polynomial arithmetic. Since all other arithmetic operations are straightforward, we will only describe how we could carry out the iterative inversion of a multivariable Hensel Code.

4.4.1 *Iterative inversion of multivariable Hensel Codes.* We now show that the iterative algorithm described in Section 5.5.6 of Chapter IV for obtaining the multiplicative inverse of a Hensel code can be extended in a straightforward manner for the multivariable case. Let us for convenience write

$$H(p,r,\alpha(x_1, x_2, \ldots, x_n))$$
$$= a(x_1, x_2, \ldots, x_n)$$
$$= \sum a_{ijkl\ldots} x_1^i x_2^j \ldots x_n^k + O(r)$$
$$= a\, O(r).$$

As before, $O(r)$ means keeping all the terms $x_1^{\beta_1} \cdots x_n^{\beta_n}$ such that $\sum_{i=1}^{n} \beta_i \leq r-1$ and rejecting all the terms beyond.

We assume that $a_{00\cdots 0} \neq 0$. The Newton–Hensel method first constructs the starting approximation

$$b_{2^i}(i = 0) = b_{2^0} = a_{0\cdots 0}^{-1} \bmod p$$

and performs the iteration

$$b_{2^i} = b_{2^{i-1}}[2 - ab_{2^{i-1}}]\, O(2^i)$$

for $i \geq 1$ until we have the required length of the result when $H(p,r,\alpha^{-1})$ is obtained.

PROOF. Similar to Theorem 5.5.7 of Chapter IV except that we replace mod x^{2^i} by $O(2^i)$.   □

4.4.2 EXAMPLE. Find $H(7,4,(1+x+2y)^{-1}$. We have

$$b_1 = b_{2^0} = a_{00}^{-1} \bmod 7 = 1$$
$$b_{2^1} = b_1(2 - ab_1)\, O(2)$$
$$= 1(2 - (1+x+2y)) = 1 + 6x + 5y$$
$$b_{2^2} = b_{2^1}[2 - (1+x+2y)b_{2^1}]\, O(2^2)$$
$$= H(7,4,(1+x+2y)^{-1}) = (1+6x+5y) \cdot [2 - (1+x+2y)(1+6x+5y)]$$
$$= 1 + 6x + 5y + x^2 + 4y^2 + 4xy + 6x^3 + 6y^3 + 2xy^2 + x^2y$$

(compare with Example 4.2.4).

4.5 Inversion of multivariable polynomial matrix— Gaussian elimination

Since all the basic arithmetic operations are defined for multivariable polynomial Hensel codes it is possible to use Gaussian elimination (with scaling) to solve linear systems with multivariable polynomial coefficients in a manner similar to that described in Section 7 of Chapter IV. As described earlier, we need to have an estimate of the coefficient size and degree of the rational polynomial elements in the inverse matrix, to use this method.

We will not describe this method here.

4.6 Iterative inversion of a multivariable polynomial matrix

The Hensel–Newton method described in Section 8 of Chapter IV can be extended to invert a nonsingular multivariable polynomial matrix. The resulting Hensel Codes of the elements can then be converted to the rational polynomials using the Padé method of solution (Section 6.2, Chapter IV).

Let $A = (a_{ij}(x_1, x_2, \ldots, x_n))$ be a nonsingular $(m \times m)$ matrix whose elements are n-variable rational polynomials over $\mathbb{I}_p$. Let us also assume that the elements of A^{-1} belong to

$$P(R-1, \ldots, R-1/R-1, \ldots, R-1, F_p(x_1, x_2, \ldots, x_n)).$$

The Newton–Schultz method first constructs

$$A \, O(1) \quad \text{and} \quad A^{-1} O(1) = B_{2^0} = B_1.$$

Then B_1 is used as the initial or starting approximation and the iteration

$$B_{2^i} = B_{2^{i-1}}[2I - AB_{2^{i-1}}] O(2^i); \quad i \geqslant 1,$$

is performed until we have the required length of the result, which is determined by the maximal degree of the elements in A^{-1} and the requirement (4.1). As before, here, $O(2^i)$ means retaining only those terms of the form $x_1^{\beta_1}, x_2^{\beta_2}, \ldots, x_n^{\beta_n}$ such that

$$\sum_{i=1}^{n} \beta_i \leqslant (2^i - 1).$$

PROOF. Use a similar argument as in Theorem 8 of Chapter IV, replacing mod by $O(\cdot)$. $\square$

4.6.1 EXAMPLE. Let

$$A = \begin{bmatrix} x+2 & y+1 \\ x+1 & y+2 \end{bmatrix} \quad \text{over } F_7(x,y)$$

$$A\,O(1) = \begin{bmatrix} 2 & 1 \\ 1 & 2 \end{bmatrix}$$

$$A^{-1}\,O(1) = \begin{bmatrix} 3 & 2 \\ 2 & 3 \end{bmatrix} = B_1$$

$$B_2 = \begin{bmatrix} 3 & 2 \\ 2 & 3 \end{bmatrix}\left[\begin{bmatrix} 2 & 0 \\ 0 & 2 \end{bmatrix} - \begin{bmatrix} x+2 & y+1 \\ x+1 & y+2 \end{bmatrix}\begin{bmatrix} 3 & 2 \\ 2 & 3 \end{bmatrix}\right]$$

$$= \begin{bmatrix} 3 & 2 \\ 2 & 3 \end{bmatrix}\left[\begin{bmatrix} 2 & 0 \\ 0 & 2 \end{bmatrix} - \begin{bmatrix} 3x+2y+1 & 2x+3y \\ 3x+2y & 2x+3y+1 \end{bmatrix}\right]$$

$$= \begin{bmatrix} 3 & 2 \\ 2 & 3 \end{bmatrix}\begin{bmatrix} 4x+5y+1 & 5x+4y \\ 4x+5y & 5x+4y+1 \end{bmatrix}$$

$$= \begin{bmatrix} 6x+4y+3 & 4x+6y+2 \\ 6x+4y+2 & 4x+6y+3 \end{bmatrix}$$

$$B_4 = B_2 \cdot \left[\begin{bmatrix} 2 & 0 \\ 0 & 2 \end{bmatrix} - \begin{bmatrix} x+2 & y+1 \\ x+1 & y+2 \end{bmatrix} \cdot B_2\right].$$

We stop the iterations here (for the sake of simplicity), since the Padé rationals are uniquely recoverable. Note, however, that according to (4.1) we need to iterate i times, until $2^i \geqslant 2R(n-1)+1 = 5$ or $i \geqslant 3$. We have now:

$$B_4 = \begin{bmatrix} (3x^3+2y^3+x^2y+5x^2+y^2+6xy+6x+4y+3) \\ (3x^3+2y^3+x^2y+5x^2+y^2+6xy+6x+4y+2) \end{bmatrix}$$

$$\begin{matrix} (2x^3+3y^3+xy^2+x^2+5y^2+6xy+4x+6y+2) \\ (2x^3+3y^3+xy^2+x^2+5y^2+6xy+4x+6y+3) \end{matrix}\end{bmatrix}.$$

Assuming that the form of the elements are

$$\begin{bmatrix} \dfrac{(a_{00}+a_{10}+a_{01}y)}{(1+b_{10}x+b_{01}y)} & \dfrac{(c_{00}+c_{10}x+c_{01}y)}{(1+d_{10}x+d_{01}y)} \\[2ex] \dfrac{(f_{00}+f_{10}x+f_{01}y)}{(1+g_{10}x+g_{01}y)} & \dfrac{(p_{00}+p_{10}x+p_{01}y)}{(1+q_{10}x+q_{01}y)} \end{bmatrix}$$

we solve for a_{00}, a_{10}, a_{01}, etc., using the Padé method (Section 6.2, Chapter IV) and obtain

$$A^{-1} = \begin{bmatrix} \dfrac{(3+5y)}{(1+5x+5y)} & \dfrac{(2+2y)}{(1+5x+5y)} \\[2ex] \dfrac{(2+2x)}{(1+5x+5y)} & \dfrac{(3+5x)}{(1+5x+5y)} \end{bmatrix}.$$

4.7 Iterative g-inversion

The iterative method described in Section 8.4 Chapter IV can be extended to multivariable polynomial matrices. Let $A = (a_{ij}(x_1, \ldots, x_n))$ be an $(m \times m)$ matrix whose elements are n-variable rational polynomial functions over $\mathbb{I}_p$. Let us assume that reflective inverse A_R^- exists and its elements belong to $P(R-1, \ldots, R-1/R-1, \ldots, R-1, F_p(x_1, x_2, \ldots, x_n))$. In the Hensel–Newton method we first construct

$$A_R^- O(1) = B_1;$$

it is assumed $A_R^- O(1)$ exists and satisfies

$$A A_R^- A\, O(1) = A$$

$$A_R^- A A_R^-\, O(1) = A_R^-.$$

We then use the iteration

$$B_{2^i} = B_{2^{i-1}}[2I - A B_{2^{i-1}}]\, O(2^i), \qquad i \geq 1$$

for $i = 1, 2, \ldots, k$, where k is determined by the maximal degree of the elements of A_R^-.

The above procedure can be extended to rational polynomials over $\mathbb{Q}$ in the usual manner described earlier, by choosing a proper prime p or a power of primes that would be adequate to represent uniquely all the elements of A_R^-.

4.7.1 EXAMPLE. Let

$$A = \begin{bmatrix} 1+x+y & 1+x+y \\ 1+x+y & 1+x+y \end{bmatrix} \in F_7(x,y)$$

$$A_R^- O(1) = \begin{bmatrix} 1 & 0 \\ 0 & 0 \end{bmatrix} = B_1$$

(see Example 7.3.3, Chapter IV). Thus

$$B_2 = \begin{bmatrix} 1 & 0 \\ 0 & 0 \end{bmatrix}\left[\begin{bmatrix} 2 & 0 \\ 0 & 2 \end{bmatrix} - \begin{bmatrix} 1+x+y & 1+x+y \\ 1+x+y & 1+x+y \end{bmatrix}\begin{bmatrix} 1 & 0 \\ 0 & 0 \end{bmatrix}\right]$$

$$= \begin{bmatrix} 1 & 0 \\ 0 & 0 \end{bmatrix}\begin{bmatrix} 1-x-y & 0 \\ -1-x-y & 2 \end{bmatrix} = \begin{bmatrix} 1-x-y & 0 \\ 0 & 0 \end{bmatrix}$$

$$B_4 = \begin{bmatrix} 1-x-y & 0 \\ 0 & 0 \end{bmatrix}\left[\begin{bmatrix} 2 & 0 \\ 0 & 2 \end{bmatrix} - \begin{bmatrix} 1-(x+y)^2 & 0 \\ 1-(x+y)^2 & 0 \end{bmatrix}\right]$$

$$= \begin{bmatrix} 1-x-y & 0 \\ 0 & 0 \end{bmatrix}\begin{bmatrix} 1+(x+y)^2 & 0 \\ -1+(x+y)^2 & 2 \end{bmatrix}$$

$$= \begin{bmatrix} 1-x-y+x^2+y^2+2xy-3x^2y-3xy^2-x^3-y^3 & 0 \\ 0 & 0 \end{bmatrix}.$$

This on conversion to Padé rational gives

$$A_R^- = \frac{1}{1+x+y}\begin{bmatrix} 1 & 0 \\ 0 & 0 \end{bmatrix}.$$

4.7.2 **Remark.** The above algorithm is quadratically convergent. One can use higher order convergent schemes as described in Section 8.1.2, Chapter IV.

EXERCISES V

1. Find $H(7,8,(3+x+y)^{-1})$ using Hensel–Newton method.

2. Find $H(7,8,(1+x+y+z)^{-1})$ using Hensel–Newton method.

3. Find the rational equivalent of

$$H(7,4,\alpha(x,y)) = 2x^3 + 3y^3 + x^2 + 5y^2 + xy^2 + 6xy + 4x + 6y + 3.$$

4. Invert the following matrix over $\mathbb{Q}(x,y)$, using Hensel–Newton method.

$$\begin{bmatrix} (x+y) & (x-y) \\ (x-y) & (x+y) \end{bmatrix}$$

5. Invert the following matrix over $F_7(x,y)$ using the direct reduction method.

$$\begin{bmatrix} x & y+1 & z+2 \\ x+2 & y+3 & z+4 \\ x+3 & y+4 & z+5 \end{bmatrix}$$

6. Invert the following matrix over $F_3(x,y)$

$$\begin{bmatrix} 1+x & 2+x \\ 2+y & 2+y \end{bmatrix}$$

using Hensel–Newton method.

7. Using a proper data structure write a general program for inverting a three-variable polynomial matrix of size (5×5) and with degrees of elements (polynomials) not exceeding three in each of the independent variables.

8. Find the Smith Canonical form of the matrix A over $\mathbb{I}$.

$$A = \begin{bmatrix} -2 & 0 & 10 \\ 0 & -3 & -4 \\ 1 & 2 & -1 \end{bmatrix}$$

9. Find the Smith Canonical form of the matrix A over $\mathbb{I}[x]$.

$$A = \begin{bmatrix} x(x-1)^3 & 0 & 0 \\ 0 & (x-1) & 0 \\ 0 & 0 & x \end{bmatrix}$$

10. Find a generalized inverse of

$$A = \begin{bmatrix} x+y & x+y \\ x-y & x-y \end{bmatrix} \text{ over } F_{17}(x,y).$$

Bibliography

Aberth, O. *Computable Analysis*, McGraw-Hill, New York, 1980.

Agarwal, R.C. and Burrus, C.S. Number Theoretic Transforms to Implement Fast Digital Convolution. *Proc. IEEE*, **63**, 1975, 550–559.

Aho, A.V., Hopcroft, J.E., and Ullman, J.D. *The Design and Analysis of Computer Algorithms*. Addison-Wesley, Reading, Mass., 1975.

Alagar, V.S. and Roy, A.K. A Comparative Study of Algorithms for Computing the Smith Normal Form of an Integer Matrix. *International J. of Systems Science* (to appear).

Albert, A.A. *Fundamental Concepts of Higher Algebra*. University of Chicago Press, Chicago, 1956.

Apostol, T.M. *Introduction to Analytic Number Theory*. Springer-Verlag, New York, 1976.

Arbib, M.A. and Manes, E.G. The Pattern of Calls Expansion is the Canonical Fixpoint for Recursive Definitions. *Jour. ACM*, **29**, 1982, 577–602.

Ax, J. and Kochen, S. Diophantine Problems over Local Fields, Parts I and II. *American J. Maths.* **87**, 1965, 605–648, and **88**, 1966, 437–456.

Bachman, G. *Introduction to p-adic Numbers and Valuation Theory*. Academic Press, New York, 1964.

Baker, G.A. *Essentials of Padé Approximants*. Academic Press, New York, 1975.

Baker, G.A. and Graves-Morris, P. Padé Approximants, Parts I and II. *Encyclopaedia of Mathematics*, Vols. 13 and 14. Addison-Wesley, Reading, Mass., 1981.

Barendregt, H.P. *The Lambda Calculus*, North-Holland, New York, 1984.

Barnett, S. *Matrices in Control Theory*. Van Nostrand Reinhold, London, 1971.

Berkelamp, E.R. *Algebraic Coding Theory*. McGraw-Hill, New York, 1968.

Bhaskara Rao, K.P.S. On Generalized Inverses of Matrices over Principal Ideal Rings. *Linear and Multilinear Algebra*, **10**, 1981, 145–154.

Bhaskara Rao, K.P.S. On Generalized Inverses over Integral Domains. *Linear Algebra and Its Applications*, **49**, 1983, 179–189.

Blum, E.K. *Numerical Analysis and Computation: Theory and Practice*. Addison-Wesley, Reading, Mass., 1972.

Borodin, A. and Munro, I. *The Computational Complexity of Algebraic and Numeric Problems*. Elsevier, New York, 1975.

Bose, N.K. *Applied Multidimensional Systems Theory*. Van Nostrand Reinhold, New York, 1983.

Brent, R.P., Gustavson, F.G., and Yun, D.Y.Y. Fast Solution of Toeplitz Systems of Equations and Computation of Padé Approximants. *Journal of Algorithms*, **1**, 1980, 259–295.

Buchberger, B., Collins, G.E., and Loos, R. *Computer Algebra—Symbolic and Algebraic Computations*. Springer-Verlag, New York, 1983.

Buchberger, B. Gröbner Bases: An Algorithmic Method in Polynomial Ideal Theory. *Recent Trends in Multidimensional Systems Theory*, ed. N.K. Bose. Reidel, Boston, 1984.

Buchberger, B. and Finkler, F. Gröbner Bases, Polynomial Remainder Sequences and Decoding of Multivariable Codes. *Recent Trends in Multidimensional Systems Theory*, ed. N.K. Bose. Reidel, Boston, 1984.

Campbell, S.L. and Meyer, C.D. *Generalized Inverses of Linear Transformations*. Pitman, London, 1979.

Campbell, S.L. *Recent Applications of Generalized Inverses*. Pitman, London, 1982.

Cherlin, G. Model Theoretical Algebra—Selected Topics. *Lecture Notes in Mathematics*, eds. Dold, A. and Eckmann, B., **521**. Springer-Verlag, New York, 1976.

Childs, L. *A Concrete Introduction to Higher Algebra*. Springer-Verlag, New York, 1979.

Chisholm, J.S.R. Rational Approximations Defined from Double Power Series. *Mathematics of Computation*, **27**, 1973, 841–848.

Chisholm, J.S.R. and McEwan, J. Rational Approximants defined from Power Series in N-variables. *Proc. Royal Society*, London, **336A**, 1974, 421–452.

Chisholm, J.S.R. N-variable Rational Approximation. *Padé and Rational Approximation—Theory and Applications*, eds. Saff, E.B. and Varga, R.S. Academic Press, New York, 1977, 23–42.

Chou, T.W. and Collins, G.E. Algorithms for Solutions of System of Linear Diophantine Equations. *SIAM J. Computing*, **11**, 1982, 687–695.

Cohen, P., Decision Procedures for real and p-adic fields. *Comm. Pure and Applied Maths.*, **22**, 1969, 131–151.

Cohn, P.M. *Universal Algebra*. Harper and Row, New York, 1965.

Cooley, J.W. and Tukey, J.W. An Algorithm for Machine Calculation of Complex Fourier Series. *Mathematics of Computation*, **19**, 1961, 297–301.

Davenport, J. Effective Mathematics—The Computer Algebra Viewpoint. *Lecture Notes in Mathematics*, eds. Dold, A, and Eckmann, B., **873**. Springer-Verlag, New York, 1981.

Davis, P.J. *Circulant Matrices*. John Wiley, New York, 1979.

de Bakker, J. *Mathematical Theory of Program Correctness*. Prentice-Hall, Englewood-Cliffs, NJ, 1980.

Delvos, P.J. and Pasdorf, H. Nth Order Blending, in Constructive Theory of Functions of Several Variables. *Lecture Notes in Mathematics*, eds. Dold, A. and Eckmann, B., **571**. Springer-Verlag, New York, 1977

Frumkin, M.A. Polynomial Time Algorithm in the Theory of Diophantine Equations. *Lecture Notes in Computer Science*, eds. Goos, G. and Hartmanis, J., **56**, 386–390. Springer-Verlag, New York, 1975.

Gantmacher, F.R. *The Theory of Matrices*, Vols. 1 and 2. Chelsea, New York, 1959.

Gold, B. and Rader, C.M. *Digital Processing of Signals*. McGraw-Hill, New York, 1969.

Gordon, W.J. Blending Function Methods of Bivariate and Multivariate Interpolation and Approximation. *SIAM J. Numerical Analysis*, **8**, 1971, 158–177.

Graham, A. *Kronecker Products and Matrix Calculus with Applications*. Ellis Horwood, Chichester, 1982.

Gregory, R.T. and Krishnamurthy, E.V. *Methods and Applications of Error-Free Computation*. Springer-Verlag, New York, 1984.

Hardy, G.H. and Wright, E.M. *An Introduction to the Theory of Numbers*. Clarendon Press, Oxford, 1979.

Hartley, P.J. Tensor Product Approximation to Data Defined in Rectangular Meshes in N-space. *Computer Journal*, **19**, 1976, 348–351.

Henrici, P. Fast Fourier Methods in Computational Complex Analysis, *SIAM Review*, **21**, 1979, 481–527.

Jacobson, N. *Lectures on Abstract Algebra*, Vols. I, II and III. Springer-Verlag, New York, 1976.

Kannan, R. and Bachem, A. Polynomial Algorithms for Computing Smith and Hermite Normal Form of an Integer Matrix. *SIAM J. Computing*, **8**, 1979, 499–507.

Kannan, R. Solving Systems of Linear Equations over Polynomials. Report CMU-CS-83-165, Department of Computer Science, Carnegie-Mellon University, Pittsburgh, 1983.

Kelley, J.L. and Namioka, I. *Linear Topological Spaces*, Springer-Verlag, New York, 1976.

Kfoury, A.J., Moll, R.N., and Arbib, M.A. *A Programming Approach to Computability*. Springer-Verlag, New York, 1982.

Kluge, W.E. Cooperative Reduction Machines. *IEEE Trans. Computers*, **32**, 1983, 1002–1012.

Knuth, D.E. *The Art of Computer Programming*, Vols. 1 and 2. Addison-Wesley, Reading, Mass., 1980.

Koblitz, N. *p-adic Numbers, p-adic Analysis, and Zeta Functions*. Springer-Verlag, New York, 1977.

Lauer, M. Generalized p-adic Constructions. *SIAM J. Computing*, **12**, 1983, 395–410.

Le Veque, W.J. *Fundamentals of Number Theory*. Addison-Wesley, Reading, Mass., 1977.

Lipson, J.D. *Elements of Algebra and Algebraic Computing*. Addison-Wesley, Reading, Mass., 1981.

Loos, R. Computing Rational Zeros of Integral Polynomials by p-adic Expansion. *SIAM J. Computing*, **12**, 1983, 286–293.

MacDuffee, C.C. *An Introduction to Abstract Algebra*. John Wiley, New York, 1948.

Mahler, K. *Lectures on Diophantine Approximations*. University of Notre Dame, 1961.

Mahler, K. *p-Adic Numbers and Their Functions*, Cambridge University Press, Cambridge, 1981.

McClellan, J.H. and Rader, C.M. *Number Theory in Digital Signal Processing*. Prentice-Hall, Englewood-Cliffs, NJ, 1980.

McClellan, M.T. The Exact Solution of Linear Equations with Polynomial Coefficients. *Jour. ACM*, **20**, 1973, 563–568.

McClellan, M.T. The Exact Solution of Linear Equations with Rational Function Coefficients. *ACM Trans. Math Software (TOMS)*, **3**, 1977, 1–25.

McEliece, R.J. The Theory of Information and Coding. *Encyclopaedia of Mathematics and Its Applications*, Vol. 2. Addison-Wesley, Reading, Mass., 1977.

McEliece, R.J. and Shearer, J.B. A Property of Euclid's Algorithm and an

Application to Padé Approximation. *SIAM J. Appl. Math.* **34,** 1978, 611–617.

McKinney, E.H. Generalized Recursive Multivariable Interpolation. *Mathematics of Computation,* **26,** 1972, 723–726.

Merz, G. Fast Fourier Transform Algorithms with Applications. *Computational Aspects of Complex Analysis,* eds. Werner, H. et al. Reidel, Boston, 1983.

Newman, M. *Integral Matrices.* Academic Press, New York, 1972.

Nussbaumer, H.J. *Fast Fourier Transform and Convolution Algorithm.* Springer-Verlag, New York, 1980.

Pereyra, V. and Scherer, G. Efficient Computer Manipulation of Tensor Products with Applications to Multidimensional Approximations. *Mathematics of Computation,* **27,** 1973, 595–605.

Pollard, J.W. The Fast Fourier Transform in a Finite Field. *Mathematics of Computation,* **25,** 1971, 365–374.

Rabiner, L.R. and Gold, B. *Theory and Application of Digital Signal Processing.* Prentice-Hall, Englewood-Cliffs, NJ, 1976.

Rabiner, L.R. and Rader, C.M. *Digital Signal Processing.* IEEE Press, New York, 1972.

Rader, C.M. Discrete Convolutions via Mersenne Transforms. *IEEE Trans. Computers,* **21,** 1972, 1269–1273.

Rao, C.R. and Mitra, S.K. *Generalized Inverse of Matrices and its Applications.* John Wiley, New York, 1971.

Reed, I.S. and Truong, T.K. The Use of Finite Fields to Compute Convolutions. *IEEE Trans. Information Theory,* **21,** 1975, 208–213.

Robinson, A. Algorithms in Algebra. *Lecture Notes in Mathematics,* eds. Dold, A. and Eckmann, B., **498,** 31–43. Springer-Verlag, New York, 1976.

Rosenbrock, H.H. and Storey, C. *Mathematics of Dynamical Systems.* Thomas Nelson, London, 1970.

Schumaker, L.L. Fitting Surfaces to Scattered Data. *Proc. Conf. Approx. Theory,* Univ. Texas, eds. Chui, G.K., Lorentz, G.G., and Schumaker, L.L. Academic Press, New York, 1977.

Scott, D. Data Types as Lattices, *SIAM J. Computing,* **5,** 1976, 522–587.

Scott, D. and Strachey, C. Toward a Mathematical Semantics for Programming Languages. PRG-6, Oxford University, 1971.

Shepperdson, J.C. and Frohlich, A. Effective Procedures in Field Theory. *Phil. Trans. Royal Society of London,* **248**A, 1956, 407–432.

Stolarsky, K.B. *Algebraic Numbers and Diophantine Approximation.* Marcel Dekker, New York, 1974.

Stoy, J.E. *The Scott–Strachey Approach to Programming Language Theory.* M.I.T. Press, Cambridge, Mass., 1979.

Tennent, R.D. *Principles of Programming Languages.* Prentice-Hall, Englewood Cliffs, NJ, 1981.

Tourlakis, G.J. *Computability.* Prentice-Hall, Englewood Cliffs, NJ, 1984.

Turner, J.A. Implementation Technique for Applicative Languages. *Software Practice and Experience,* **9,** 1979, 31–49.

Winograd, S. On Computing the Discrete Fourier Transform. *Mathematics of Computation,* **32,** 1978, 175–199.

Young, D.M. and Gregory, R.T. *A Survey of Numerical Mathematics,* Vols. I and II. Addison-Wesley, Reading, Mass., 1972.

Yun, D.Y.Y. Algebraic Algorithms using p-adic Constructions. *Proc. ACM Symp. Symbolic and Algebraic Computation* (*SYMSAC76*), ed. Jenks, R. 1976, 246–259.

Yun, D.Y.Y. On the use of a Class of Algebraic Mapping Techniques for Some

Problems in Complex Analysis. *Computational Aspects of Complex Analysis,* eds. Werner, H. et al. Reidel, Boston, 1983.

Zariski, O. and Samuel, P. *Commutative Algebra*, Vols. I and II, Springer-Verlag, New York, 1975.

Zassenhaus, H. On Hensel Factorization. *Journal of Number Theory,* **1,** 1969, 291–311.

Index

Texts and Monographs in Computer Science

continued

Arto Salomaa and Matti Soittola
Automata-Theoretic Aspects of Formal Power Series
1978. x, 171pp. cloth

Jeffrey R. Sampson
Adaptive Information Processing: An Introductory Survey
1976. x, 214pp. 83 illus. cloth

William M. Waite and Gerhard Goos
Compiler Construction
1984. xiv, 446pp. 196 illus. cloth

Niklaus Wirth
Programming in Modula-2
3rd Corr. Edition. 1985. iv, 208pp. cloth